CCNA™
Routing and Switching Practice Tests

Robert Gradante

CCNA™ Routing and Switching Practice Tests Exam Cram

Limits Of Liability And Disclaimer Of Warranty

Trademarks

The Coriolis Group, LLC
14455 N. Hayden Road
Suite 220
Scottsdale, Arizona 85260

480/483-0192
FAX 480/483-0193
http://www.coriolis.com

Library of Congress Cataloging-in-Publication Data
Gradante, Robert
    CCNA routing and switching practice tests exam cram/by Bob Gradante.
        p.   cm.
    Includes index.
    ISBN 1-57610-542-3
    1. Electronic data processing personnel--Certification.   2.
Telecommunication--Switching systems--Examinations--Study guides.   I. Title.
QA76.3.G75   2000
004.6'6--dc21   99-049139
                                                                     CIP

President, CEO
Keith Weiskamp

Publisher
Steve Sayre

Acquisitions Editor
Jeff Kellum

Marketing Specialist
Cynthia Caldwell

Project Editor
Stephanie Palenque

Technical Reviewer
Donald Fuller

Production Coordinator
Wendy Littley

Cover Design
Jesse Dunn

Layout Design
April Nielsen

CD-ROM Developer
Michelle McConnell

Printed in the United States of America
10 9 8 7 6 5 4 3 2 1

CORIOLIS™

The Coriolis Group, LLC • 14455 North Hayden Road, Suite 220 • Scottsdale, Arizona 85260

ExamCram.com Connects You to the Ultimate Study Center!

Our goal has always been to provide you with the best study tools on the planet to help you achieve your certification in record time. Time is so valuable these days that none of us can afford to waste a second of it, especially when it comes to exam preparation.

Over the past few years, we've created an extensive line of *Exam Cram* and *Exam Prep* study guides, practice exams, audio training, and interactive training. To help you study even better, we have now created an e-learning and certification destination called **ExamCram.com**. (You can access the site at **www.examcram.com**.) Now, with every study product you purchase from us, you'll be connected to a large community of people like yourself who are actively studying for their certifications, developing their careers, seeking advice, and sharing their insights and stories.

I believe that the future is all about collaborative learning. Our **ExamCram.com** destination is our approach to creating a highly interactive, easily accessible collaborative environment, where you can take practice exams and discuss your experiences with others, sign up for features like "Questions of the Day," plan your certifications using our interactive planners, create your own personal study pages, and keep up with all of the latest study tips and techniques.

I hope that whatever study products you purchase from us—*Exam Cram* or *Exam Prep* study guides, *Practice Tests, Flash Cards, Personal Trainers*, or one of our interactive Web courses—that our products will make your studying fun and productive. Our commitment is to build the kind of learning tools that will allow you to study the way you want to, whenever you want to.

Help us continue to provide the very best certification study materials possible. Write us or email us at **learn@examcram.com** and let us know

Visit ExamCram.com now to enhance your study program.

how our study products have helped you study. Tell us about new features that you'd like us to add. Send us a story about how we've helped you. We're listening!

Good luck with your certification exam and your career. Thank you for allowing us to help you achieve your goals.

Keith Weiskamp
President and CEO

Look For These Other Books From The Coriolis Group:

CCNA Routing and Switching Exam Cram, Second Edition
Jeffrey T. Coe, Matthew J. Rees, and Jason Waters

CCNA Routing and Switching Exam Prep
Mark Poplar, Jason Waters, Shawn McNutt, and David Stabenaw

CCNA Routing and Switching Exam Cram Personal Trainer
CIP Author Team

CCNA Routing and Switching Exam Cram Flashcards
Robert Gradante

CCNA Routing and Switching Exam Cram Audio Review
Jason Waters

CCNA Routing and Switching Exam Cram Value Pack
CIP Author Team

For my family, the most important people in my life.

About The Author

Robert Gradante works as the technical training manager of New Horizons Computer Learning Center of Long Island. He currently holds CCNA,MCSE+I, A+, Network +, INET +, and MCT certifications. Robert has published several titles on networking.

Acknowledgments

I am very excited about the way this book turned out, and I must give all the credit to the people at The Coriolis Group for making this happen. I'd like to thank Jeff Kellum for giving me the opportunity to once again contribute to the fine line of technical titles in the Coriolis family, and for surrounding me with good people to get the job done. Special thanks to Stephanie Palenque, my project editor, who did a marvelous job of coordinating all phases of this project, and was always a pleasure to deal with. Thanks also to the rest of the Coriolis team who worked on this book: Paula Kmetz, Wendy Littley, April Nielsen, and Jesse Dunn. Many thanks to Bonnie Smith for another excellent copyediting job. Bonnie, you are a true professional. Thanks also to my technical reviewer, Donald Fuller. I could not have asked for a more competent and knowledgeable person to review this material. Thank you Don, for adding depth and clarity to this work.

Table Of Contents

Introduction

You Spoke

Welcome to *CCNA Routing and Switching Practice Tests Exam Cram*! A recent survey of Coriolis readers showed us how important practice questions are in your efforts to prepare to take and pass certification exams. You asked us to give you more practice tests on a variety of certification topics, including CCNA, MCSE Core Four, MCSE+I, A+, Network+, and others.

We Responded

The *Practice Tests Exam Cram* series is our answer to your requests, and provides you with entirely new practice tests for many certification topics. Each practice test appears in its own chapter, followed by a corresponding answer and explanation chapter, in the same format as the Sample Test and Answer Key chapters at the end of each of our *Exam Cram* books. We not only tell you which answers are correct, but we also explain why the right answers are right and why the wrong answers are wrong. That's because we're convinced that you can learn as much from understanding the wrong answers as you can from knowing the right ones!

This book makes a perfect companion to any study material you may own that covers the exam subject matter. For those of you who already own the *CCNA Routing and Switching Exam Cram, Second Edition* book, we have included a time-saving study feature. At the end of each answer, you will find a reference to the *Exam Cram* book or to another valuable resource in which the question topic is discussed most thoroughly. That way, if you want to review the material on which the question is based in more depth, you will be able to quickly locate that information.

But Wait, There's More!

This book also includes a CD-ROM that contains one more exam.. This additional exam is built using an interactive format that allows you to practice in an exam environment similar to Cisco's own testing format.

Thus, this book gives you access to a pool of 270 questions. Thorough review of these materials should provide you with a reasonably complete view of the numerous topics and types of questions you're likely to see on a real Cisco exam. Because questions come and go on Cisco exams pretty regularly, we can't claim total coverage, but we have designed these question pools to deal with the topics and concepts that are most likely to appear on a real exam in some form or fashion.

Using This Book To Prepare For An Exam

You should begin your preparation process by working through the materials in the book to guide your studies. As you discover topics or concepts that may be unfamiliar or unclear, be sure to consult additional study materials to increase your knowledge and familiarity with the materials involved. In fact, you should employ this particular technique on any practice test questions you come across that may expose areas in your knowledge base that may need further development or elaboration.

To help you increase your knowledge base, we suggest that you work with whatever materials you have at hand. Certainly, we can't help but recommend our own *Exam Cram* and *Exam Prep* books, but you will find that the *Exam Crams* also cite numerous other sources of information as well.

Among those tools, some other materials worth obtaining include any training kits available from Cisco Press for software under discussion. For this book, that clearly means the *Cisco CCNA Preparation Library* and *CCNA Exam Certification Guide*. You can learn more about these and other Cisco Press titles online at **http://www.ciscopress.com**. Once you've worked your way through the text-based practice tests in the book, use the interactive exam on the CD to assess your test readiness. That way, you can build confidence in your ability to sit for and pass these exams, as you master the subject material for each one.

We recommend you shoot for a passing result when you take the practice tests. If you don't make the grade, you probably should take some other practice exams—possibly from one or more of the vendors shown below, all of whom offer practice exams for under $100 (in most cases, well under $100). The following are some vendors of practice exams and their corresponding URLs:

➤ Network Study Guides—**www.networkstudyguides.com**

➤ Free Networking Encyclopedia online—**http://www.webopedia.com**

Use these practice exams and resources to increase your knowledge, familiarity, and understanding of the materials involved in the topic area, rather than spending the money on a Cisco test that you may not pass. Because each Cisco exam costs $100, we think you're better off spending some of that money on more preparation, and saving what you can to help pay for the exam when you're really ready to pass it!

Tell Us What You Think

Feel free to share your feedback on the book with us. We'll carefully consider your comments. Please be sure to include the title of the book in your message; otherwise, we'll be forced to guess which book you are writing about. Please send your comments and questions to us at **learn@examcram.com**.

Visit our Web site at **www.examcram.com** for the latest on what's happening in the world of certification, updates, and new *Exam Prep* and *Exam Cram* titles. For the latest information on Cisco certification exams, visit Cisco's Web site at **www.cisco.com**. Good luck with your exams!

Practice Test #1

Question 1

Seated at your router console, you want to view the startup configuration that the router is using. Which of the following is the correct Cisco operating system (IOS) command used to display the contents of the startup configuration of a Cisco router?

○ a. **ROUTER> SHOW STARTUP-CONFIG**

○ b. **ROUTER#SHOW STARTUP-CONFIG**

○ c. **ROUTER>SHOW STARTUP CONFIG**

○ d. **ROUTER#SHOW RUNNING CONFIG**

Question 2

When two computers are communicating over a network, they routinely exchange service requests and responses with each other. Which layer of the Open Systems Interconnection (OSI) model is responsible for coordinating the service requests and responses that occur between these computers across the network?

○ a. Application

○ b. Presentation

○ c. Session

○ d. Transport

○ e. Network

○ f. Data link

○ g. Physical

Question 3

Consider the following display on your router console:

```
Router con0 is now available

Press RETURN to get started.

User Access Verification

Password:
```

Based on this display, what type of password must be entered at the Password prompt in order to continue?

○ a. You must enter the console password to continue.

○ b. You must enter the virtual terminal password to continue.

○ c. You must enter the auxiliary password to continue.

○ d. You must enter the enable password to continue.

Question 4

Which of the following statements is true about hierarchical and flat addressing?

- ○ a. The flat addressing used in IPX provides a more organized environment than hierarchical addressing.

- ○ b. The hierarchical addressing used in IPX provides a more organized environment than flat addressing.

- ○ c. The flat addressing of IP addresses enables complex networks to be logically grouped and organized.

- ○ d. The hierarchical addressing of data link addresses allows complex networks to be logically grouped and organized.

Question 5

You have just made changes to your Cisco router's running configuration and wish to save this to the configuration that is stored in NVRAM. Which of the following is the correct command to perform this operation?

- ○ a. **COPY STARTUP-CONFIG RUNNING-CONFIG**

- ○ b. **COPY RUNNING-CONFIG STARTUP-CONFIG**

- ○ c. **COPY RUNNING-CONFIG TFTP**

- ○ d. **COPY RUNNING-CONFIG TO STARTUP CONFIG**

Question 6

Seated at your Cisco router, you wish to rerun the initial System Configuration Dialog. You issue the **SETUP** command, and receive the following output:

```
Router>SETUP
Translating "setup"...
domain server (255.255.255.255)
% Unknown command or computer name,
or unable to find computer address
```

Which of the following best describes why this operation failed?

○ a. You cannot issue the **SETUP** command in user mode. The **SETUP** command must be run from privileged mode.

○ b. You cannot issue the **SETUP** command in privileged mode. The **SETUP** command must be run from user mode.

○ c. You issued the wrong command. The correct command to rerun the initial System Configuration Dialog is **INSTALL**.

○ d. The operation did not fail; this is the expected output of the **SETUP** command.

Question 7

You are configuring your Cisco router and need to enter configuration commands from the terminal. Which of the following is the correct command and prompt used to enter the **CONFIGURE TERMINAL** command?

○ a. **ROUTER> CONFIG T**

○ b. **ROUTER#CONFIG T**

○ c. **ROUTER(config)#CONFIG T**

○ d. **ROUTER(CONFIG-IF)#CONFIG T**

Question 8

As the network administrator, you want to require a password to be entered by any user connecting to your Cisco router through the auxiliary port. You want the password to be "KULA". Which of the following are the correct Cisco IOS commands to perform this operation?

◯ a.
```
AUX 0 4
   LOGIN
   PASSWORD KULA
```

◯ b.
```
LINE CON 0 4
   LOGIN
   PASSWORD KULA
```

◯ c.
```
LINE VTY 0 4
   LOGIN KULA
```

◯ d.
```
LINE AUX 0
   LOGIN
   PASSWORD KULA
```

Question 9

Key components of WAN communication are DTE and DCE devices. Which of the following best describes a DCE device?

○ a.   DCE or device connection equipment are modems used for communicating over digital phone lines.

○ b.   DCE or data circuit-terminating equipment are the devices between the LAN and the DTE that perform the data translation between the LAN and the DTE (data terminal equipment).

○ c.   DCE or data carrier equipment is usually the router that acts as the gateway to the LAN.

○ d.   DCE or data carrier equipment are devices owned by the telephone company.

Question 10

You are configuring IP on serial interface s1 of your Cisco router. You specify the following at the router console:

```
Configuring interface Serial1:
Is this interface in use? [no]: y
Configure IP on this interface? [no]: y
Configure IP unnumbered on
this interface? [no]: n
IP address for this
interface: 200.200.200.254
Number of bits in
subnet field [0]: 2
```

Based on this information, which of the following statements is true?

○ a.   This is a Class C address, and the subnet mask is 255.255.255.224.

○ b.   This is a Class C address, and the subnet mask is 255.255.255.192.

○ c.   This is a Class B address, and the subnet mask is 255.255.255.248.

○ d.   This is a Class C address, and the subnet mask is 255.255.255.0.

Question 11

Which of the following correctly states the use of the up arrow at the command line in a Cisco router?

- ○ a. The up arrow key will move the cursor to the beginning of the command line.

- ○ b. The up arrow key will recall previous commands in the command history.

- ○ c. The up arrow key will move the cursor back one word.

- ○ d. The up arrow key will return to more recent commands in the command history after recalling previous commands.

Question 12

You have been assigned a Class C IP address for use in your network. You will use subnet mask 255.255.255.240.

Based on this information, how many total bits are available for host IDs on this network?

- ○ a. 16

- ○ b. 6

- ○ c. 4

- ○ d. 15

Question 13

Which layer of the OSI model utilizes Abstract Syntax Notation One (ASN 1) to negotiate data transfer syntax?

- ○ a. Application

- ○ b. Presentation

- ○ c. Session

- ○ d. Transport

- ○ e. Network

- ○ f. Data link

- ○ g. Physical

Question 14

Which of the following is the correct key(s) used to return to the beginning of the command line in a Cisco router?

○ a. Ctrl+A

○ b. Ctrl+E

○ c. Ctrl+P

○ d. Ctrl+N

○ e. Esc+B

○ f. Down arrow

○ g. Up arrow

Question 15

Which of the following best describes the use of the auxiliary password in a Cisco router?

○ a. The auxiliary password is used to control access to the Ethernet port on the router.

○ b. The auxiliary password is used to control remote Telnet access to a router.

○ c. The auxiliary password is used to control access to any auxiliary ports the router may have.

○ d. The auxiliary password is used to control access to the privileged mode of the router.

○ e. The auxiliary password is used to control access to the console port of the router.

○ f. The auxiliary password is used to control access to the user mode of the router.

Question 16

Which of the following are networking standards for sending data over a physical medium? [Choose the three best answers]

- ❑ a. FDDI
- ❑ b. Token Ring
- ❑ c. Ethernet
- ❑ d. SLIP
- ❑ e. PPP

Question 17

Which of the following statements are true about connection-oriented and connectionless protocols? [Choose the two best answers]

- ❑ a. Connectionless protocols feature the creation of a virtual circuit.
- ❑ b. UDP is a connection-oriented protocol.
- ❑ c. UDP is a connectionless protocol.
- ❑ d. Connectionless protocols' overhead is less than connection-oriented protocols.

Question 18

Reliable end-to-end communication is the primary function of the transport layer of the OSI model. A number of protocols exist at the transport layer that provide these services. Which of the following are transport layer protocols?

- ○ a. SMTP and Telnet are transport layer protocols.
- ○ b. SPX and TCP are transport layer protocols.
- ○ c. IP and IPX are transport layer protocols.
- ○ d. SQL and RPC are transport layer protocols.

Question 19

The OSI model defines standards used in networking and comprises a seven-layer model. Which layer of the OSI model acts as a "window" to network resources for applications?

○ a. Application

○ b. Presentation

○ c. Session

○ d. Transport

○ e. Network

○ f. Data link

○ g. Physical

Question 20

Integrated Services Digital Network (ISDN) is a communications standard that uses digital telephone lines to transmit voice, data, and video. ISDN is a dial-up service that is used on demand, and ISDN protocol standards are defined by the International Telecommunications Union (ITU). Which of the following statements is true about ISDN protocol standards?

○ a. Protocols that begin with the letter *Q* define ISDN standards on the existing telephone network.

○ b. Protocols that begin with the letter *I* define ISDN standards on the existing telephone network.

○ c. Protocols that begin with the letter *Q* specify ISDN switching and signaling standards.

○ d. Protocols that begin with the letter *E* specify ISDN concepts, interfaces, and terminology.

Question 21

All of the protocols that are used in networking conform to one of the seven layers of the OSI model. Which of the following statements is true about protocols and the OSI model?

○ a. SMTP, FTP, and EDI are all presentation layer protocols.

○ b. FTP, Telnet, and TIFF are all application layer protocols.

○ c. Telnet, WAIS, and MIDI are all presentation layer protocols.

○ d. MPEG, PICT, and MIDI are all presentation layer protocols.

Question 22

The bandwidth provided by ISDN can be divided into two categories: Primary Rate ISDN (PRI) and Basic Rate ISDN (BRI). Which of the following statements is true about Primary Rate ISDN?

○ a. Primary Rate ISDN (also known as 23B+1D) comprises 23 B channels that are 64kbps each and one 16kbps D channel.

○ b. Primary Rate ISDN (also known as 23B+1D) comprises 23 B channels that are 64kbps each and one 64kbps D channel.

○ c. Primary Rate ISDN (also known as 23D+1B) comprises 23 D channels that are 64kbps each and one 64kbps B channel.

○ d. Primary Rate ISDN (also known as 23B+2D) comprises 23 B channels that are 64kbps each and two 64kbps D channels.

Question 23

Which of the following core components of a Cisco router is used as the router's main working area and contains the running configuration?

○ a. Flash

○ b. RAM

○ c. ROM

○ d. NVRAM

Question 24

Which of the following access lists will allow all hosts on network 195.20.200.0 to have FTP access to network 151.120.0.0?

○ a. **ACCESS-LIST 102 PERMIT TCP 151.120.0.0 0.0.255.255 195.20.200.0 .0.0.0.255 EQ 21**

○ b. **ACCESS-LIST 101 PERMIT TCP 195.20.200.0 0.0.0.255 151.120.0.0 0.0.255.255 EQ 21**

○ c. **ACCESS-LIST 110 PERMIT TCP 151.120.0.0 .0.0.255.255 195.20.200.0 0.0.0.255 EQ 21**

○ d. **ACCESS-LIST 101 PERMIT TCP 195.20.200.0 0.0.0.255 151.120.0.0 0.0.255.255 EQ 23**

Question 25

Frame relay networks utilize Local Management Interface (LMI) signaling types, which are used to manage a frame relay connection. Which of the following is the Cisco default LMI signaling type used in frame relay networks?

○ a. Q933A

○ b. ANSI

○ c. CISCO

○ d. HDLC

Question 26

You are configuring your Cisco router to use RIP. Which of the following is the correct command to enter enable RIP on the router and have it advertise network 200.200.200.0?

○ a.
```
ROUTER RIP 200.200.200.0
```

○ b.
```
NETWORK RIP
    NETWORK 200.200.200.0
```

○ c.
```
ROUTER-RIP
    NETWORK 200.200.200.0
```

○ d.
```
ROUTER RIP
    NETWORK 200.200.200.0
```

Question 27

A Cisco router's startup sequence is divided into three basic operations. What is the first step in this startup sequence?

○ a. The router's configuration file is loaded into memory from NVRAM.

○ b. The bootstrap program in ROM executes the Power On Self Test (POST).

○ c. The Cisco IOS (operating systems software) is loaded into memory from ROM.

○ d. The Cisco IOS (operating systems software) is loaded into memory per instructions from the boot system command.

Question 28

You are monitoring the matches that are made to IP access list number 155 on your Cisco router. After making some changes in the network, you decide to check the matches to this access list again. Before you take these new readings, you would like to clear the previous matches from access list 155. Which of the following Cisco IOS commands could you use to perform this operation?

○ a. **CLEAR ACCESS-LIST-COUNTERS 155**

○ b. **CLEAR ACCESS-LIST MATCHES 155**

○ c. **CLEAR ACCESS-LIST COUNTERS 155**

○ d. **CLEAR ACCESS LIST COUNTERS 155**

Question 29

You manage a medium-sized network that utilizes the Internetwork Packet Exchange/Sequenced Packet Exchange (IPX/SPX) protocol. The network is segmented by a Cisco router. To control the amount of traffic that passes through the router, you decide to use a standard IPX access list. You want to create an access list that will deny IPX network 20 access to IPX network 40, but you want all other networks to be able to access network 40. Which of the following is the correct access list to use in this situation?

○ a. **ACCESS-LIST 800 DENY 20 40**

   ACCESS-LIST 800 PERMIT –1 –1

○ b. **ACCESS-LIST 800 DENY 40 20**

   ACCESS-LIST 800 PERMIT –1 –1

○ c. **ACCESS LIST 800 DENY 20 40**

   ACCESS-LIST 100 PERMIT –1 –1

○ d. **IPX ACCESS-LIST 800 DENY 20 40**

   ACCESS-LIST 800 PERMIT –1 –1

Question 30

You manage the Cisco routers for your company. Your company maintains a total of seven routers. After modifying the running configuration on one of these routers, you save the running configuration to the startup configuration. Soon after you do this, you realize that you have configured the wrong router. All of the new settings that you have changed on this router must be removed. Fortunately, you do have a backup copy of the original configuration file on a remote TFTP server. Which of the following commands could you use to return this router to the configuration it was using before you made these changes?

○ a. Issue the **COPY TFTP STARTUP CONFIG** command and restart the router.

○ b. Issue the **COPY TFTP STARTUP-CONFIG** command and restart the router.

○ c. Issue the **COPY STARTUP-CONFIG TFTP** command and restart the router.

○ d. Issue the **COPY STARTUP CONFIG TFTP** command and restart the router.

Question 31

You have configured several IP access lists on your Cisco router. Which of the following commands can you use to see a list of all the IP access lists that are currently in use on interface S0 of your Cisco router?

- ○ a. **SHOW IP ACCESS-LISTS S0**
- ○ b. **SHOW IP INTERFACE S0**
- ○ c. **SHOW IP ACCESS-LISTS**
- ○ d. **SHOW IP INTERFACE-ACCESS-LISTS S0**

Question 32

Which of the following statements is true about the protocols that exist at the network layer of the OSI model?

- ○ a. SMTP and Telnet are network layer protocols.
- ○ b. SPX and TCP are network layer protocols.
- ○ c. IP and IPX are network layer protocols.
- ○ d. SQL and RPC are network layer protocols.

Question 33

The OSI model defines a layered approach to network communication. Each layer of the OSI model adds its own layer-specific information and passes it on to the next layer, until it leaves the computer and goes out onto the network. This process is known as encapsulation. Encapsulation involves five steps. First, user information is converted to data. What is the second step of encapsulation?

- ○ a. Data is converted to packets.
- ○ b. Data is converted to datagrams.
- ○ c. Data is converted to segments.
- ○ d. Data is converted to bits.

Question 34

When two computers are communicating over a network, which layer of the OSI model is responsible for ensuring the reliable transfer of data by implementing connection-oriented services between nodes?

○ a. Application

○ b. Presentation

○ c. Session

○ d. Transport

○ e. Network

○ f. Data link

○ g. Physical

Question 35

The Cisco IOS includes a command history feature that can be used to store previously entered commands in a history buffer. Which of the following commands would you enter to disable the command history feature?

○ a. **NO TERMINAL EDITING**

○ b. **TERMINAL NO EDITING**

○ c. **DISABLE TERMINAL EDITING**

○ d. **NO EDITING**

Question 36

You manage a medium-sized network that utilizes the IPX/SPX protocol. The network is segmented by a Cisco router. To control the amount of traffic that passes through the router, you decide to use an extended IPX access list. You want to create an access list that will deny IPX network 44 access to IPX network 41, but you want all other networks to be able to access network 41. Which of the following is the correct access list to use in this situation?

- ○ a. **ACCESS LIST 900 DENY –1 44 0 41 0**

  ACCESS-LIST 900 PERMIT –1 –1 0 –1 0

- ○ b. **ACCESS-LIST 900 DENY –1 41 0 44 0**

  ACCESS-LIST 900 PERMIT –1 –1 0 –1 0

- ○ c. **ACCESS-LIST 900 DENY –1 44 0 41 0**

  ACCESS-LIST 900 PERMIT –1 –1 0 –1 0

- ○ d. **ACCESS-LIST 800 DENY –1 44 0 41 0**

  ACCESS-LIST 800 PERMIT –1 –1 0 –1 0

Question 37

Both RIP and IGRP are dynamic routing protocols used to automatically update entries in a routing table. Which of the following statements are true about IGRP and RIP? [Choose the three best answers]

- ❑ a. RIP routers will broadcast the contents of their routing table every 30 seconds.
- ❑ b. RIP has a maximum hop count of 255.
- ❑ c. IGRP considers hop count, delay, bandwidth, and reliability when determining the best path.
- ❑ d. RIP is a distance vector protocol that is proprietary to Cisco.
- ❑ e. RIP is a link state protocol that is proprietary to Cisco.
- ❑ f. Both IGRP and RIP are link state routing protocols.
- ❑ g. Any destination that requires more than 15 hops is considered unreachable by RIP.

Question 38

Flow control is a method used by TCP at the transport layer that ensures the amount of data being sent from one node will not overwhelm the destination node. What flow control method predetermines the amount of data that will be sent before an acknowledgment is expected?

○ a. Windowing

○ b. Buffering

○ c. Source-quench messages

○ d. Multiplexing

Question 39

Which of the following Exec modes contains the **CONFIGURE** command that is used to access other configuration modes, such as global configuration and interface modes?

○ a. User mode contains the **CONFIGURE** command.

○ b. Privileged mode contains the **CONFIGURE** command.

○ c. Both user mode and privileged mode contain the **CONFIGURE** command.

○ d. Administrative mode contains the **CONFIGURE** command.

Question 40

The Cisco IOS includes a context-sensitive help feature that can assist an operator when entering commands into a Cisco router. Which of the following is the correct command to use when you would like to see a list of all available commands for a particular router mode?

○ a. **?**

○ b. **HELP <COMMAND>**

○ c. **LIST COMMANDS**

○ d. **SHOW COMMANDS**

Question 41

You need to configure an extended IP access list on your Cisco router. You need to control users' access by port. Which of the following is not a valid entry for the port field in an IP extended access list?

○ a. FTP

○ b. SMTP

○ c. HTTP

○ d. TELNET

○ e. DNS

○ f. A port number from 0 to 65,535

○ g. IP

Question 42

Your network contains a Netware file server. The Netware file server's address is 8E.1234.1234.1234. You do not want your Cisco router to pass any Service Advertising Protocol (SAP) advertisements relating to file services from this server to other networks. To filter this traffic, you decide to create a SAP access list. Which of the following is the correct IPX SAP access list used to perform this operation?

○ a. **ACCESS LIST 1000 DENY 8E.1234.1234.1234 4**

○ b. **ACCESS-LIST 999 DENY 8E.1234.1234.1234 4**

○ c. **ACCESS-LIST 1000 DENY 8E.1234.1234.1234 7**

○ d. **ACCESS-LIST 1000 DENY 8E.1234.1234.1234 4**

Question 43

The most common form of WAN (wide area network) connectivity utilizes telecommunication technology. Which of the following best describes the DEMARC, as it is used in WAN communication?

○ a. The DEMARC is the collection of communication devices (telephones, modems, and so forth) that exists at the customer's location.

○ b. The DEMARC is the telephone company's office that acts as the central communication point for the customer.

○ c. The DEMARC is the cabling that extends from the CPE to the telephone company's office.

○ d. The DEMARC is the point at the customer's premises where CPE devices connect and is usually a large punch-down board located in a wiring closet.

Question 44

A switch is a data link layer connectivity device used to segment a network. Cisco switches employ two basic methods of forwarding packets: cut-through and store-and-forward. Which of the following statements are correct about cut-through and store-and-forward switching? [Choose the two best answers]

❏ a. Store-and-forward switches will copy an incoming packet to the local buffer and perform error checking on the packet before sending it on to its destination.

❏ b. Cut-through switches will copy an incoming packet to the local buffer and perform error checking on the packet before sending it on to its destination.

❏ c. Store-and-forward switching experiences less latency than cut-through switching.

❏ d. Cut-through switching experiences less latency than store-and-forward switches.

Question 45

> Which of the following protocols does not reside at the data link layer of the OSI model?
>
> ○ a. DNA SCP
>
> ○ b. SDLC
>
> ○ c. X.25
>
> ○ d. SLIP
>
> ○ e. PPP
>
> ○ f. ISDN
>
> ○ g. LAPB

Question 46

> Which of the following best describes the function and contents of Random Access Memory (RAM) in a Cisco router?
>
> ○ a. RAM is a physical chip installed on a router's motherboard that contains the bootstrap program, the POST (Power On Self Test), and the operating system software (Cisco IOS).
>
> ○ b. RAM is used by Cisco routers to store the startup configuration.
>
> ○ c. RAM is an erasable, programmable memory area that contains the Cisco operating system software (IOS).
>
> ○ d. RAM is used as the router's main working area and contains the running configuration.

Question 47

You are configuring a welcome message to be displayed by your Cisco router to users when they log in. Which of the following is the Cisco IOS command used to configure a Cisco router to display a logon banner that displays the message "Welcome to my router" when users log in?

○ a.
```
BANNER MOTD #
    Welcome to my router #
```

○ b.
```
SET LOGON BANNER
    Welcome to my router
```

○ c.
```
BANNER #
    Welcome to my router #
```

○ d.
```
SET WELCOME MESSAGE #
    Welcome to my router #
```

Question 48

You are troubleshooting a Cisco router and need to display the configuration information of port Ethernet 0. Which of the following Cisco IOS commands could you issue to view this information?

○ a. **DISPLAY ETHERNET 0**

○ b. **SHOW ETHERNET 0**

○ c. **SHOW INTERFACE ETHERNET 0**

○ d. **DISPLAY INTERFACE ETHERNET 0**

Question 49

Which of the following is the correct numerical range used for an extended IPX access list?

○ a. 1–99

○ b. 100–199

○ c. 800–899

○ d. 900–999

Question 50

You are configuring a standard IP access list and enter the following at the router prompt:

```
ACCESS LIST 10 PERMIT 200.200.200.0 255.255.255.0
```

Based on this information, which of the following statements best describes the result of this access list?

○ a. This access list will allow traffic from host 200.200.200.0.

○ b. This access list will allow traffic from host 200.200.200.0 to enter network 255.255.255.0.

○ c. This access list will allow traffic from all hosts on network 200.200.200.0.

○ d. This access list is invalid.

Question 51

Consider the following IPX address:

7g.9999.8888.7777

Based on this information, what is the node ID of this address?

○ a. 9999.8888.7777

○ b. 7g.9999

○ c. 7g.9999.8888

○ d. 7g

Question 52

Routing refers to the process of determining the path a packet will take to its destination. A Cisco router cannot route anything unless there is an entry in its routing table that contains information on where to send the packet. Which of the following statements are true about the methods used to place entries in a routing table? [Choose the two best answers]

❏ a. Dynamic routing protocols, such as RIP and IGRP, can automatically update a routing table.

❏ b. Routed protocols, such as IP and IPX, can update a routing table.

❏ c. Static routes can be used to update a routing table.

❏ d. The **UPDATE** command can be used to update a routing table.

Question 53

A problem that can occur with distance vector routing protocols are routing loops. Distance vector routing protocols utilize a number of different methods to prevent routing loops. Which of the following are methods used by distance vector routing protocols to prevent routing loops? [Choose the four best answers]

❏ a. New horizon

❏ b. Route poisoning

❏ c. Hold-downs

❏ d. Maximum hop count

❏ e. Split horizon

❏ f. Horizon poisoning

Question 54

You manage a network that uses TCP/IP as the only protocol and is divided into multiple subnets. The IP configuration of one of your subnets is as follows:

```
Broadcast Address = 192.23.98.63 on interface E0
Subnet Mask = 255.255.255.240
```

Based on this information, what are the valid host IDs for the nodes in this subnet?

- ○ a.   192.23.98.32 to 192.23.98.63
- ○ b.   192.23.98.49 to 192.23.98.62
- ○ c.   192.23.98.48 to 192.23.98.63
- ○ d.   192.23.98.33 to 192.23.98.62

Question 55

Which of the following statements best describes the use of the **CONFIGURE TERMINAL (CONFIG T)** command?

- ○ a.   **CONFIGURE TERMINAL** is used to enter configuration commands into the router from the console port or through Telnet.
- ○ b.   **CONFIGURE TERMINAL** is used to copy the configuration file from a TFTP server into the router's RAM.
- ○ c.   **CONFIGURE TERMINAL** is used to copy a configuration file into NVRAM from a TFTP server.
- ○ d.   **CONFIGURE TERMINAL** is used to execute the configuration stored in NVRAM and will copy the startup configuration to the running configuration.

Question 56

In an effort to provide security on your Cisco router, you want to require a password to access the privileged mode of the router. You decide to use the enable password, and you want the password to be "Swordfish". Which of the following is the correct Cisco IOS command used to set the enable password to "Swordfish" on a Cisco router?

○ a. **ENABLE PASSWORD Swordfish**

○ b. **MAKE PASSWORD Swordfish**

○ c. **LINE ENABLE Swordfish**

○ d. **ENABLE SWORDFISH**

Question 57

The most common types of connections used in WANs can be divided into digital communication, analog communication, and packet switching technology. Which of the following WAN technology describes a standard analog phone line and is used as a dial-up service using modems?

○ a. ISDN

○ b. POTS

○ c. Frame relay

○ d. X.25

Question 58

You want to enable dynamic routing on your Cisco router and decide to use IGRP. Which of the following are the correct commands to enable IGRP on the router and have the router advertise network 223.20.200.0 in autonomous system number 11?

○ a.
```
ROUTER IGRP
    NETWORK 223.20.200.0 11
```

○ b.
```
ROUTER IGRP 11
    NETWORK 223.20.200.0
```

○ c.
```
ROUTER-IGRP 11
    NETWORK 223.20.200.0
```

○ d.
```
ROUTER IGRP AS 11
    NETWORK 223.20.200.0
```

Question 59

Which of the following core components of a Cisco router cannot permanently store information after the router is powered off?

○ a. Flash

○ b. RAM

○ c. ROM

○ d. NVRAM

Question 60

What are the two primary modes used by the Cisco command interpreter EXEC? [Choose the two best answers]

❏ a. User mode

❏ b. Privileged mode

❏ c. Executive mode

❏ d. Settings mode

Question 61

Consider the following display on your Cisco router:

```
Router con0 is now available
Press RETURN to get started.
Router>ENABLE
Router#
```

What mode is this router presently in?

○ a. User mode

○ b. Privileged mode

○ c. Executive mode

○ d. Global Configuration mode

Question 62

Which of the following Cisco IOS commands is used to enter configuration commands into the router from the console port or through Telnet?

○ a. **CONFIGURE TERMINAL (CONFIG T)**

○ b. **CONFIGURE OVERWRITE-NETWORK (CONFIG O)**

○ c. **CONFIGURE NETWORK (CONFIG NET)**

○ d. **CONFIGURE MEMORY (CONFIG MEM)**

Question 63

To control the amount of SAP advertisement traffic that is passed by your Cisco router, you create the following IPX SAP access list:

```
ACCESS-LIST 1080 DENY 8E.1234.1234.1234 7
ACCESS-LIST 1080 DENY 5f.5656.5656.5656 4
ACCESS LIST 1080 PERMIT -1
```

You want to apply this access list to the outgoing serial interface number 0 of your router. You enter the interface configuration mode of your router and access serial interface 0.

Which one of the following commands will you enter to perform this operation?

○ a. **IPX OUTPUT-SAP-FILTER 1080**

○ b. **OUTPUT-SAP-FILTER 1080**

○ c. **IPX OUTPUT SAP FILTER 1080**

○ d. **IPX ACCESS-GROUP 1080 OUT**

Question 64

A dynamic routing table is one that has its entries entered automatically by the router, through the use of a routing protocol. A dynamic routing protocol provides for automatic discovery of routes. Common categories of routing protocols are link state and distance vector. Which of the following statements is true about distance vector and link state routing protocols?

○ a. RIP is a link state routing protocol.

○ b. OSPF is a distance vector routing protocol.

○ c. EIGRP is a link state routing protocol.

○ d. IGRP is a distance vector routing protocol.

○ e. IP is a link state routing protocol.

○ f. IPX is a distance vector routing protocol.

Question 65

The TCP/IP configuration of one of the computers in your network is as follows:

```
IP address 131.200.50.90
Subnet Mask 255.255.224.0
```

Based on this information, how many bits *in total* are being used by subnet mask 255.255.224.0?

- ○ a. 16
- ○ b. 19
- ○ c. 20
- ○ d. 18

Question 66

Which of the following best describes the use of the **HELP** command in a Cisco router?

- ○ a. The **HELP** command can be used to obtain command-line syntax help when used with a command (HELP COMMAND).
- ○ b. The **HELP** command is not a valid Cisco IOS command.
- ○ c. The **HELP** command is used to provide basic instructions on how to use the context-sensitive Help feature of Cisco routers.
- ○ d. The **HELP** command is used to open the graphical Cisco Help feature that can be used to obtain command-line help.

Question 67

Which of the following correctly states the use of Ctrl+A at the command line in a Cisco router?

- ○ a. Ctrl+A will move the cursor to the beginning of the command line.
- ○ b. Ctrl+A will move the cursor to the end of the command line.
- ○ c. Ctrl+A will move the cursor back one word.
- ○ d. Ctrl+A will move the cursor forward one word.

Question 68

You are experiencing problems with the startup configuration on your Cisco router. You have a backup copy of this configuration stored on a TFTP server. Which of the following is the correct Cisco IOS command that you could use to copy the configuration file from your TFTP server to the startup configuration of your router?

○ a. **COPY TFTP STARTUP CONFIG**

○ b. **COPY STARTUP-CONFIG TFTP**

○ c. **COPY TFTP-STARTUP-CONFIG**

○ d. **COPY TFTP STARTUP-CONFIG**

Question 69

You need to change the Host name of your Cisco router. The new host name will be "ROUTERA". Which of the following is the correct command you will enter at your router console to make this change?

○ a. **HOSTNAME ROUTERA**

○ b. **SET HOSTNAME ROUTERA**

○ c. **HOST ROUTERA**

○ d. **ASSIGN HOSTNAME ROUTERA**

Question 70

Access lists can be used to filter the traffic that is handled by a Cisco router. Which of the following can be specified in an extended IP access list?

○ a. Protocol

○ b. Port

○ c. Source

○ d. Destination

○ e. All of the above

Answer Key #1

1. b	19. a	37. a, c, g	54. b
2. c	20. c	38. a	55. a
3. a	21. d	39. b	56. a
4. b	22. b	40. a	57. b
5. b	23. b	41. g	58. b
6. a	24. b	42. d	59. b
7. b	25. c	43. d	60. a, b
8. d	26. d	44. a, d	61. b
9. b	27. b	45. a	62. a
10. b	28. c	46. d	63. a
11. b	29. a	47. a	64. d
12. c	30. b	48. c	65. b
13. b	31. b	49. d	66. c
14. a	32. c	50. d	67. a
15. c	33. c	51. a	68. d
16. a, b, c	34. d	52. a, c	69. a
17. c, d	35. b	53. b, c, d, e	70. e
18. b	36. c		

Question 1

The correct answer is b. The **SHOW STARTUP-CONFIG** command can be used in privileged mode to display the startup configuration file stored in nonvolatile RAM (NVRAM). Answer a is incorrect, because the router is in user mode, and the **SHOW STARTUP-CONFIG** command must be executed from privileged mode. In user mode, the router prompt is followed by an angle bracket (**ROUTER>**). In privileged mode, the router prompt is followed by a pound sign (**ROUTER#**). Answer c is incorrect, because the **SHOW STARTUP-CONFIG** command has a hyphen between **STARTUP** and **CONFIG**. Answer d is incorrect, because the **SHOW RUNNING-CONFIG** command is used to display the current running configuration in RAM, not the startup configuration.

For more information, see Chapter 8 of *CCNA Routing and Switching Exam Cram, Second Edition.*

Question 2

The correct answer is c. The Open Systems Interconnection (OSI) model defines standards used in networking and comprises a seven-layer model. The session layer of the OSI model is primarily used to coordinate service requests and responses that occur between nodes across the network, as well as creating, maintaining, and ending a communication session. Based on this, all of the other answers are incorrect.

The following list outlines the seven layers of the OSI model and their primary functions:

➤ *Application*—The "window" to networking used by programs; the application layer is responsible for:

  ➤ Verifying that the appropriate resources are present to initiate a connection with the destination node.

  ➤ Verifying the identity of the destination node.

Application layer protocols include Telnet, File Transfer Protocol (FTP), Simple Mail Transfer Protocol (SMTP), World Wide Web (WWW), Electronic Data Interchange (EDI), and Wide Area Information Server (WAIS).

➤ *Presentation*—Essentially a translator, the presentation layer is responsible for:

  ➤ Translating text and data syntax, such as Extended Binary-Coded Decimal Interchange Code (EBCDIC, used in IBM systems) and

American Standard Code for Information Interchange (ASCII, used in PC and most computer systems).

➤ Using Abstract Syntax Notation to perform data translation.

Presentation layer protocols include PICTure (PICT, Apple computer picture format), Tagged Image File Format (TIFF), Joint Photographic Experts Group (JPEG), Musical Instrument Digital Interface (MIDI), Motion Picture Expert Group (MPEG), and Quick Time (audio/video application).

➤ *Session*—Used to coordinate communication between nodes, the session layer is responsible for:

➤ Creating, maintaining, and ending a communication session.

➤ Coordinating service requests and responses that occur between nodes across the network.

Session layer protocols include Network File System (NFS, used by SUN Microsystems and Unix with TCP/IP), structured query language (SQL, used to define database information requests), remote procedure calls (RPC, used in Microsoft network communication), X Window (used by Unix terminals), AppleTalk Session Protocol (ASP, used by Apple computers), and Digital Network Architecture Session Control Protocol (DNA SCP, used by IBM).

➤ *Transport*—Reliable end-to-end communication is the primary function of the transport layer. Its many responsibilities include:

➤ Ensuring flow control so the amount of data being sent from one node will not overwhelm the destination node.

➤ Ensuring the ability of multiple applications to utilize a single transport (multiplexing).

➤ Ensuring the reliable transfer of data. The transport layer implements connection-oriented services between nodes, which utilize a three-way handshake (synchronization, acknowledgment, and data transfer) to efficiently transfer data.

➤ Ensuring positive acknowledgment or the process of a node waiting for an acknowledgment from the destination node prior to sending data.

➤ Windowing, a form of flow control, that specifies how much data will be transferred between acknowledgments.

Transport layer protocols include Transmission Control Protocol (TCP, part of the TCP/IP protocol suite), and Sequenced Packet Exchange (SPX, part of the IPX/SPX protocol suite).

➤ *Network*—Selecting the appropriate path a packet should take to get to its intended destination is the function of the network layer. It is responsible for routing, which is the process of using a network layer address to determine the best path a packet will travel to its destination. Network layer protocols include Internet Protocol (IP, part of the TCP/IP protocol suite), and Internetwork Packet Exchange (IPX, part of the IPX/SPX protocol suite).

➤ *Data Link*—Divided into two sublayers (media access control [MAC] and logical link control [LLC]), the data-link layer is responsible for:

➤ Preparing data from upper layers to be transmitted over the physical medium by encapsulating upper-layer data into frames. This frame includes the source and destination of MAC addresses in the frame header.

➤ Converting data into bits, so it can be transmitted by the physical layer.

➤ Adding a cyclical redundancy check (CRC) to the end of a frame, which is used for error checking at the data-link layer.

Data-link protocols include High-Level Data Link Control (HDLC, the Cisco default encapsulation for serial connections), Synchronous Data Link Control, (SDLC, used in IBM networks), Link Access Procedure Balanced (LAPB, used with X.25), X.25 (a packet switching network), Serial Line Internet Protocol (SLIP, an older TCP/IP dial-up protocol), Point-to-Point (PPP, a newer dial-up protocol), Integrated Services Digital Network (ISDN, a dial-up digital service), and frame relay (a packet switching network).

➤ *Physical*—The lowest layer of the OSI model, the physical layer, defines the electrical functionality required to send and receive bits over a given physical medium. Specifications that define the voltage levels and physical components of a network are at the physical layer. Protocols are not specified at the physical layer, because they are implemented as software. Examples of the standards for sending data over the physical medium are Ethernet (the most widely used standard in networking), Token Ring (IBM's proprietary networking topology), and Fiber Distributed Data Interface (FDDI, a standard for fiber optic networks, commonly used as a backbone).

For more information, see Chapter 4 of *CCNA Routing and Switching Exam Cram, Second Edition*.

Question 3

The correct answer is a. The display shows the router is at line **con0**, which is the console. The message, **User Access Verification**, signifies that a console password has been set for access. The console password is used to control access to the console port of the router.

Answer b is incorrect, because the virtual terminal password is used to control remote Telnet access to a router. Setting the virtual terminal password will require all users that Telnet into the router to provide this password for access. Answer c is incorrect, because the auxiliary password is used to control access to any auxiliary ports the router may have. Setting the auxiliary password will require all users connecting to the auxiliary port of the router (usually remote dial-in) to provide this password for access. Answer d is incorrect, because the enable password is used to control access to the privileged mode of the router. If an enable password is specified, it must be entered after the **ENABLE** command to successfully access privileged mode.

For more information, see Chapter 4 of *CCNA Routing and Switching Exam Cram, Second Edition.*

Question 4

The correct answer is b. A network address, or logical address, is an address that resides at the network layer of the OSI model. Network addresses are hierarchical in nature. A hierarchical addressing scheme uses logically structured addresses to provide a more organized environment. Examples of hierarchical network addresses are IP and IPX. An IP address is composed of a network ID, which identifies the network a host belongs to, and a host ID, which is unique to that host. An IPX address utilizes a similar network ID and host ID format in addressing. The hierarchical addressing of IP and IPX enable complex networks to be logically grouped and organized.

A data-link address is also known as a physical address, hardware address, or, more commonly, a MAC address. (So named a MAC address, because this address resides at the MAC or media access control sublayer of the data-link layer.) The MAC address is the addressed "burned" into every network adapter card by the manufacturer. This address is considered a "flat" address, because no logical arrangement of these addresses are on a network. A flat addressing scheme simply gives each member a unique identifier that is associated with him or her. Answers a and c are incorrect, because IP and IPX are hierarchical addresses, not flat addresses. Answer d is incorrect, because data-link addresses are flat addresses, not hierarchical addresses.

For more information, see Chapter 4 of *CCNA Routing and Switching Exam Cram, Second Edition.*

Question 5

The correct answer is b. **COPY RUNNING-CONFIG STARTUP-CONFIG** will copy the running configuration from RAM to the startup configuration file in NVRAM. Answer a is incorrect, because the **COPY STARTUP-CONFIG RUNNING-CONFIG** command will copy the startup configuration to the running configuration. Cisco IOS commands generally follow the source first, then the destination when using copy commands. Answer c is incorrect, because the **COPY RUNNING-CONFIG TFTP** will copy the running configuration from RAM to a remote TFTP server. Answer d is incorrect, because **COPY RUNNING-CONFIG TO STARTUP-CONFIG** is not a valid Cisco IOS command. The startup configuration is a configuration file in a Cisco router that contains the commands used to set router-specific parameters. The startup configuration is stored in NVRAM and is loaded into RAM when the router starts up.

The running configuration is the startup configuration that is loaded into RAM and is the configuration used by the router when it is running. Because any information in RAM is lost when the router is powered off, the running configuration can be saved to the startup configuration, which is located in NVRAM, and will, therefore, be saved when the router is powered off. In addition, both the running and startup configuration files can be copied to and from a TFTP server for backup purposes. The following list outlines the common commands used when working with configuration files:

➤ *COPY STARTUP-CONFIG RUNNING-CONFIG*—This will copy the startup configuration file in NVRAM to the running configuration in RAM.

➤ *COPY STARTUP-CONFIG TFTP*—This will copy the startup configuration file in NVRAM to a remote TFTP server.

➤ *COPY RUNNING-CONFIG STARTUP-CONFIG*—This will copy the running configuration from RAM to the startup configuration file in NVRAM.

➤ *COPY RUNNING-CONFIG TFTP*—This will copy the running configuration from RAM to a remote TFTP server.

➤ *COPY TFTP RUNNING-CONFIG*—This will copy a configuration file from a TFTP server to the routers running configuration in RAM.

➤ *COPY TFTP STARTUP-CONFIG*—This will copy a configuration file from a TFTP server to the startup configuration in NVRAM.

Note that the basic syntax of the **COPY** command specifies the source first, then the destination.

To modify a router's running configuration, you must be in global configuration mode. The **CONFIG T** command is used in privileged mode to access global configuration mode, which is displayed at the router prompt with "**config**" in parentheses (**RouterA(config)#**). To exit global configuration mode, use the keys, Control and Z (Ctrl+Z).

For more information, see Chapter 4 of *CCNA Routing and Switching Exam Cram, Second Edition*.

Question 6

The correct answer is a. The **SETUP** command must be issued from privileged mode, and in this question, the router is in user mode. In user mode, the router prompt is followed by an angle bracket (**ROUTER>**). The **SETUP** command will execute the System Configuration Dialog, which can be used to modify the configuration used by the router.

Answer b is incorrect, because the **SETUP** command must be run from privileged mode. In privileged mode, the router prompt is followed by a pound sign (**ROUTER#**). Answer c is incorrect, because **INSTALL** is not a valid Cisco IOS command. Answer d is incorrect, because the operation did fail, as noted by the **Translating "setup"...domain server (255.255.255.255) % Unknown command or computer name, or unable to find computer address** response from the router.

For more information, see Chapter 8 of *CCNA Routing and Switching Exam Cram, Second Edition*.

Question 7

The correct answer is b. The **CONFIGURE TERMINAL** command can be abbreviated as **CONFIG T** and must be entered from privileged mode. In privileged mode, the router prompt is followed by a pound sign (**ROUTER#**).

Answer a is incorrect, because the router is in user mode, and the **CONFIGURE TERMINAL** command cannot be entered from this mode. In user mode, the router prompt is followed by an angle bracket (**ROUTER>**). Answer c is incorrect, because the prompt is already in configuration mode. After entering configuration mode, the router prompt displays **"config"** in parentheses (**ROUTER(config)#**). Answer d is incorrect, because the prompt is displaying interface mode (**ROUTER(CONFIG-IF)#**).

For more information, see Chapter 4 of *CCNA Routing and Switching Exam Cram, Second Edition.*

Question 8

The correct answer is d. The auxiliary password is used to control access to any auxiliary ports the router may have. Setting the auxiliary password will require all users connecting to the auxiliary port of the router (usually remote dial-in) to provide this password for access. To configure the auxiliary password to **KULA,** use the following commands:

```
LINE AUX 0
LOGIN
PASSWORD KULA
```

Answer a is incorrect, because the correct syntax for setting the **AUX** password is **LINE AUX 0 4.** Answer b is incorrect, because the **LINE CON** command will configure a password on the console port, not the auxiliary port. Answer c is incorrect, because the **LOGIN** command does not specify the password, the **PASSWORD** command does.

Cisco routers utilize passwords to provide security access to a router. Passwords can be set for controlling access to privileged mode, remote sessions via Telnet, or through the auxiliary port. The following list details the common passwords used in Cisco routers:

➤ *Enable Password*—The enable password is used to control access to the privileged mode of the router. If an enable password is specified, it must be entered after the **ENABLE** command to successfully access privileged mode. To configure the enable password, use the **ENABLE PASSWORD [password]** command.

➤ *Enable Secret Password*—The enable secret password is used to control access to privileged mode, similar to the enable password. The difference between enable secret and enable password is that the enable secret password will be encrypted for additional security. The enable secret password will take precedence over the enable password if both are enabled. To configure the enable secret password, use the **ENABLE SECRET [password]** command.

➤ *Virtual Terminal Password*—The virtual terminal password is used to control remote Telnet access to a router. Setting the virtual terminal password will require all users that Telnet into the router to provide this password for access. To configure the virtual terminal password, use the following commands:

```
LINE VTY 0 4
LOGIN
PASSWORD [password]
```

➤ *Auxiliary Password*—The auxiliary password is used to control access to any auxiliary ports the router may have. Setting the auxiliary password will require all users connecting to the auxiliary port of the router (usually remote dial-in) to provide this password for access. To configure the auxiliary password, use the following commands:

```
LINE AUX 0
LOGIN
PASSWORD [password]
```

➤ *Console Password*—The console password is used to control access to the console port of the router. Setting the console password will require all users connecting to the router console to provide this password for access. To configure the console password, use the following commands:

```
LINE CON 0
LOGIN
PASSWORD [password]
```

For more information, see Chapter 8 of *CCNA Routing and Switching Exam Cram, Second Edition.*

Question 9

The correct answer is b. Key components of wide area network (WAN) communication are data terminal equipment (DTE) and data circuit-terminating equipment (DCE) devices. A DTE is usually the router that acts as the gateway to the local area network. DCE are the devices between the local area network (LAN) and the DTE that perform the data translation between the LAN and the DTE. Common DCE devices are modems (used when communicating over analog phone lines) and a channel service unit/data service unit

(CSU/DSU), used when communicating over digital lines. All of the other answers are incorrect, because they all describe the incorrect definition of a DCE device.

For more information, see Chapter 11 of *CCNA Routing and Switching Exam Cram, Second Edition.*

Question 10

The correct answer is b. When configuring IP on a Cisco router interface, you must specify the IP address for the interface and the number of bits used in the subnet mask. When an IP address is specified for the interface, the router will identify the address by the first octet rule as either a Class A, B, or C address. (Class A=1–126, Class B=128–191, and Class C=192–223.) The default subnet mask for the address class is assumed to be present by the router. (Class A=255.0.0.0, Class B=255.255.0.0, and Class C=255.255.255.0.)

The second step to configuring IP on a router interface involves specifying the number of bits in the subnet field. If this number is 0, the default subnet mask is applied based on the address class. The number of bits specified in the subnet field will be applied by the router to the host portion of the IP address. In this question, a Class C address was entered for the interface (200.200.200.254). The router then assumed a subnet mask of 255.255.255.0, the default subnet mask for a Class C address. The number of bits in the subnet field was specified as two, which was interpreted by the router as two bits in the *fourth* octet of the IP address, and is applied as 255.255.255.192. (Two bits in binary=11000000, which is the decimal equivalent to 192.) Answer a is incorrect, because subnet mask 255.255.255.224 uses three bits in the subnet mask field in a Class C address. Answer c is incorrect, because IP address 200.200.200.254 is a Class C address, not a Class B address. Answer d is incorrect, because subnet mask 255.255.255.0 uses zero bits in the subnet mask field in a class C address.

For more information, see Chapter 8 of *CCNA Routing and Switching Exam Cram, Second Edition.*

Question 11

The correct answer is b. Pressing the up arrow key will recall previous commands in the command history. (Ctrl+P will do this as well.) Answer a is incorrect, because Ctrl+A will move the cursor to the beginning of the command line. Answer c is incorrect, because Esc+B will move the cursor back one word. Answer d is incorrect, because Ctrl+N will return to more recent commands in the command history after recalling previous commands.

Configuring a Cisco router can sometimes involve long, detailed command lines that can be slow to navigate. For this reason, the Cisco IOS includes a number of "shortcut" ways to navigate the command line, and the following list details some of the more common methods to use:

➤ *Ctrl+A*—Pressing the Control key along with the letter *A* will move the cursor to the beginning of the command line.

➤ *Ctrl+E*—Pressing the Control key along with the letter *E* will move the cursor to the end of the command line.

➤ *Ctrl+P*—Pressing the Control key along with the letter *P* will recall previous commands in the command history. (The up arrow will do this as well.)

➤ *Ctrl+N*—Pressing the Control key along with the letter *N* will return to more recent commands in the command history after recalling previous commands. (The down arrow will do this as well.)

➤ *Esc+B*—Pressing the Escape key along with the letter *B* will move the cursor back one word.

➤ *Esc+F*—Pressing the escape key along with the letter *F* will move the cursor forward one word.

➤ *Left and Right Arrow Keys*—The left and right arrow keys will move the cursor one character left and right, respectively.

➤ *Tab*—Pressing the Tab key will complete an entry typed at the router prompt.

For more information, see Chapter 8 of *CCNA Routing and Switching Exam Cram, Second Edition*.

Question 12

The correct answer is c. Because the subnet mask, 255.255.255.240, is using 28 bits and because an IP address is a total of 32 bits, 4 bits remain to create host IDs on this network. An IP address is a 32-bit network layer address that is composed of a network ID and a host ID. A subnet mask is used to distinguish the network portion of an IP address from the host portion. The 32 bits in an IP address are expressed in dotted decimal notation in four octets (four 8-bit numbers separated by periods, that is, 208.100.50.90). The subnet mask, 255.255.255.240, is said to use 28 bits, because it takes 28 of 32 bit positions to create. This subnet mask, 255.255.255.240, can be expressed in binary as 11111111.11111111.11111111.11110000, which is three octets using eight bits each (24) and four bits used in the fourth octet to make the number 240. Because there are only four bits remaining to create hosts with this subnet mask, all of the other answers are incorrect.

For more information, see Chapter 8 of *CCNA Routing and Switching Exam Cram, Second Edition*.

Question 13

The correct answer is b. The OSI model defines standards used in networking and is composed of a seven-layer model. The presentation layer of the OSI model is primarily used as a "translator" for the application layer and uses Abstract Syntax Notation One (ASN 1) to negotiate or translate information to the application layer. Because the Presentation layer is the only layer of the OSI model that uses ASN 1, all of the other answers are incorrect. The following list outlines the seven layers of the OSI model and their primary functions:

➤ *Application*—The "window" to networking used by programs; the application layer is responsible for:

➤ Verifying that the appropriate resources are present to initiate a connection with the destination node.

➤ Verifying the identity of the destination node.

Application layer protocols include Telnet, File Transfer Protocol (FTP), Simple Mail Transfer Protocol (SMTP), World Wide Web (WWW), Electronic Data Interchange (EDI), and Wide Area Information Server (WAIS).

➤ *Presentation*—Essentially a translator, the presentation layer is responsible for:

➤ Translating text and data syntax, such as Extended Binary-Coded Decimal Interchange Code (EBCDIC, used in IBM systems) and American Standard Code for Information Interchange (ASCII, used in PC and most computer systems).

➤ Using Abstract Syntax Notation to perform data translation.

Presentation layer protocols include PICTure (PICT, Apple computer picture format), Tagged Image File Format (TIFF), Joint Photographic Experts Group (JPEG), Musical Instrument Digital Interface (MIDI), Motion Picture Expert Group (MPEG), and Quick Time (audio/video application).

➤ *Session*—Used to coordinate communication between nodes, the session layer is responsible for:

➤ Creating, maintaining, and ending a communication session.

➤ Coordinating service requests and responses that occur between nodes across the network.

Session layer protocols include Network File System (NFS, used by SUN Microsystems and Unix with TCP/IP), structured query language (SQL, used to define database information requests), remote procedure calls (RPC, used in Microsoft network communication), X Window (used by Unix terminals), AppleTalk Session Protocol (ASP, used by Apple computers), and Digital Network Architecture Session Control Protocol (DNA SCP, used by IBM).

➤ *Transport*—Reliable end-to-end communication is the primary function of the transport layer. Its many responsibilities include:

➤ Ensuring flow control so the amount of data being sent from one node will not overwhelm the destination node.

➤ Ensuring the ability of multiple applications to utilize a single transport (multiplexing).

➤ Ensuring the reliable transfer of data. The transport layer implements connection-oriented services between nodes, which utilize a three-way handshake (synchronization, acknowledgment, and data transfer) to efficiently transfer data.

➤ Ensuring positive acknowledgment or the process of a node waiting for an acknowledgment from the destination node prior to sending data.

➤ Windowing, a form of flow control, that specifies how much data will be transferred between acknowledgments.

Transport layer protocols include Transmission Control Protocol (TCP, part of the TCP/IP protocol suite), and Sequenced Packet Exchange (SPX, part of the IPX/SPX protocol suite).

➤ *Network*—Selecting the appropriate path a packet should take to get to its intended destination is the function of the network layer. It is responsible for routing, which is the process of using a network layer address to determine the best path a packet will travel to its destination. Network layer protocols include Internet Protocol (IP, part of the TCP/IP protocol suite), and Internetwork Packet Exchange (IPX, part of the IPX/SPX protocol suite).

➤ *Data Link*—Divided into two sublayers (media access control [MAC] and logical link control [LLC]), the data-link layer is responsible for:

➤ Preparing data from upper layers to be transmitted over the physical medium by encapsulating upper-layer data into frames. This frame includes the source and destination of MAC addresses in the frame header.

➤ Converting data into bits, so it can be transmitted by the physical layer.

➤ Adding a cyclical redundancy check (CRC) to the end of a frame, which is used for error checking at the data-link layer.

Data-link protocols include High-Level Data Link Control (HDLC, the Cisco default encapsulation for serial connections), Synchronous Data Link Control, (SDLC, used in IBM networks), Link Access Procedure Balanced (LAPB, used with X.25), X.25 (a packet switching network), Serial Line Internet Protocol (SLIP, an older TCP/IP dial-up protocol), Point-to-Point (PPP, a newer dial-up protocol), Integrated Services Digital Network (ISDN, a dial-up digital service), and frame relay (a packet switching network).

➤ *Physical*—The lowest layer of the OSI model, the physical layer, defines the electrical functionality required to send and receive bits over a given physical medium. Specifications that define the voltage levels and physical components of a network are at the physical layer. Protocols are not specified at the physical layer, because they are implemented as software. Examples of the standards for sending data over the physical medium are Ethernet (the most widely used standard in networking), Token Ring (IBM's proprietary networking topology), and Fiber Distributed Data Interface (FDDI, a standard for fiber optic networks, commonly used as a backbone).

For more information, see Chapter 4 of *CCNA Routing and Switching Exam Cram, Second Edition*.

Question 14

The correct answer is a. Pressing the Control key along with the letter *A* will move the cursor to the beginning of the command line. Configuring a Cisco router can sometimes involve long, detailed command lines that can be slow to navigate. For this reason, the Cisco IOS includes a number of "shortcuts" to navigate the command line. The following list details some of the more common methods to use:

➤ *Ctrl+A*—Pressing the Control key along with the letter *A* will move the cursor to the beginning of the command line.

➤ *Ctrl+E*—Pressing the Control key along with the letter *E* will move the cursor to the end of the command line.

➤ *Ctrl+P*—Pressing the Control key along with the letter *P* will recall previous commands in the command history. (The up arrow will do this as well.)

➤ *Ctrl+N*—Pressing the Control key along with the letter *N* will return to more recent commands in the command history after recalling previous commands. (The down arrow will do this as well.)

➤ *Esc+B*—Pressing the Escape key along with the letter *B* will move the cursor back one word.

➤ *Esc+F*—Pressing the Escape key along with the letter *F* will move the cursor forward one word.

➤ *Left and Right Arrow Keys*—The left and right arrow keys will move the cursor one character left and right, respectively.

➤ *Tab*—Pressing the Tab key will complete an entry typed at the router prompt.

For more information, see Chapter 8 of *CCNA Routing and Switching Exam Cram, Second Edition.*

Question 15

The correct answer is c. The auxiliary password is used to control access to any auxiliary ports the router may have. Cisco routers utilize passwords to provide security access to a router. Passwords can be set for controlling access to privileged mode, remote sessions via Telnet, or through the auxiliary port. Answer a is incorrect, because the console password is used to control access to the Ethernet port of the router. Answer b is incorrect, because Virtual terminal password is used to control remote Telnet access to the router. Answer d is incorrect, because the enable password is used to control access to the privileged mode of the router. Answer e is incorrect, because the console password is used to control access to the console port of the router. Answer f is incorrect, because there is no password used to enter user mode. The following list details the common passwords used in Cisco routers:

➤ *Enable Password*—The enable password is used to control access to the privileged mode of the router. If an enable password is specified, it must be entered after the **ENABLE** command to successfully access privileged mode. To configure the enable password, use the **ENABLE PASSWORD [password]** command.

➤ *Enable Secret Password*—The enable secret password is used to control access to privileged mode, similar to the enable password. The difference between enable secret and enable password is that the enable secret password will be encrypted for additional security. The enable secret password will take precedence over the enable password if both are enabled. To configure the enable secret password, use the **ENABLE SECRET [password]** command.

➤ *Virtual Terminal Password*—The virtual terminal password is used to control remote Telnet access to a router. Setting the virtual terminal password will require all users that Telnet into the router to provide this password for access. To configure the virtual terminal password, use the following commands:

```
LINE VTY 0 4
LOGIN
PASSWORD [password]
```

➤ *Auxiliary Password*—The auxiliary password is used to control access to any auxiliary ports the router may have. Setting the auxiliary password will require all users connecting to the auxiliary port of the router (usually remote dial-in) to provide this password for access. To configure the auxiliary password, use the following commands:

```
LINE AUX 0
LOGIN
PASSWORD [password]
```

➤ *Console Password*—The console password is used to control access to the console port of the router. Setting the console password will require all users connecting to the router console to provide this password for access. To configure the console password, use the following commands:

```
LINE CON 0
LOGIN
PASSWORD [password]
```

For more information, see Chapter 8 of *CCNA Routing and Switching Exam Cram, Second Edition.*

Question 16

The correct answers are a, b, and c. Standards for sending data over the physical medium are Ethernet (the most widely used standard in networking), Token Ring (IBM's proprietary networking topology), and FDDI, a standard for fiber optic networks, commonly used as a backbone. Answers d and e are incorrect, because SLIP, an older TCP/IP dial-up protocol, and PPP, a newer dial-up protocol, are dial-up protocols that reside at the data-link layer of the OSI model.

For more information, see Chapter 4 of *CCNA Routing and Switching Exam Cram, Second Edition*.

Question 17

The correct answers are c and d. User Datagram Protocol (UDP) is a transport layer protocol that is connectionless. Answer a is incorrect, because connectionless protocols do not utilize virtual circuits. Answer b is incorrect, because UDP is a connectionless protocol, not a connection-oriented protocol. Connectionless protocols feature:

➤ *No Sequencing or Virtual Circuit Creation*—Connectionless protocols do not sequence packets or create virtual circuits.

➤ *No Guarantee of Delivery*—Packets are sent as datagrams, and delivery is not guaranteed by connectionless protocols. When using a connectionless protocol like UDP, the guarantee of delivery is the responsibility of higher layer protocols.

➤ *Less Overhead*—Because connectionless protocols do not perform any of the above services, overhead is less than connection-oriented protocols. Transmission Control Protocol (TCP) is a protocol that is connection-oriented. Connection-oriented protocols feature:

➤ Reliability when a communication session is established between hosts before sending data. This session is considered a virtual circuit.

➤ Sequencing, because when a connection-oriented protocol like TCP sends data, it numbers each segment, so the destination host can receive the data in the proper order.

➤ Guaranteed delivery of connection-oriented protocols utilize error checking to guarantee packet delivery.

➤ More overhead because of the additional error checking and sequencing responsibilities, connection-oriented protocols require more overhead than connectionless protocols.

For more information, see Chapter 11 of *CCNA Routing and Switching Exam Cram, Second Edition.*

Question 18

The correct answer is b. Transport layer protocols include TCP, part of the TCP/IP protocol suite, and SPX, part of the IPX/SPX protocol suite. Answer a is incorrect, because SMTP and Telnet are application layer protocols. Answer c is incorrect, because IP and IPX are network layer protocols. Choice d is incorrect, because SQL and RPC are session layer protocols.

For more information, see Chapter 9 of *CCNA Routing and Switching Exam Cram, Second Edition.*

Question 19

The correct answer is a. The application layer of the OSI model functions as the "window" for applications to access network resources and is responsible for verifying that the appropriate resources exist to make a connection. Since no other layer of the OSI model performs this function, all of the other answers are incorrect.

The following list outlines the seven layers of the OSI model and their primary functions:

➤ *Application*—The "window" to networking used by programs; the application layer is responsible for:

➤ Verifying that the appropriate resources are present to initiate a connection with the destination node.

➤ Verifying the identity of the destination node.

Application layer protocols include Telnet, File Transfer Protocol (FTP), Simple Mail Transfer Protocol (SMTP), World Wide Web (WWW), Electronic Data Interchange (EDI), and Wide Area Information Server (WAIS).

➤ *Presentation*—Essentially a translator, the presentation layer is responsible for:

➤ Translating text and data syntax, such as Extended Binary-Coded Decimal Interchange Code (EBCDIC, used in IBM systems) and American Standard Code for Information Interchange (ASCII, used in PC and most computer systems).

➤ Using Abstract Syntax Notation to perform data translation.

Presentation layer protocols include PICTure (PICT, Apple computer picture format), Tagged Image File Format (TIFF), Joint Photographic Experts Group (JPEG), Musical Instrument Digital Interface (MIDI), Motion Picture Expert Group (MPEG), and Quick Time (audio/video application).

➤ *Session*—Used to coordinate communication between nodes, the session layer is responsible for:

➤ Creating, maintaining, and ending a communication session.

➤ Coordinating service requests and responses that occur between nodes across the network.

Session layer protocols include Network File System (NFS, used by SUN Microsystems and Unix with TCP/IP), structured query language (SQL, used to define database information requests), remote procedure calls (RPC, used in Microsoft network communication), X Window (used by Unix terminals), AppleTalk Session Protocol (ASP, used by Apple computers), and Digital Network Architecture Session Control Protocol (DNA SCP, used by IBM).

➤ *Transport*—Reliable end-to-end communication is the primary function of the transport layer. Its many responsibilities include:

➤ Ensuring flow control so the amount of data being sent from one node will not overwhelm the destination node.

➤ Ensuring the ability of multiple applications to utilize a single transport (multiplexing).

➤ Ensuring the reliable transfer of data. The transport layer implements connection-oriented services between nodes, which utilize a three-way handshake (synchronization, acknowledgment, and data transfer) to efficiently transfer data.

➤ Ensuring positive acknowledgment or the process of a node waiting for an acknowledgment from the destination node prior to sending data.

➤ Windowing, a form of flow control, that specifies how much data will be transferred between acknowledgments.

Transport layer protocols include Transmission Control Protocol (TCP, part of the TCP/IP protocol suite), and Sequenced Packet Exchange (SPX, part of the IPX/SPX protocol suite).

➤ *Network*—Selecting the appropriate path a packet should take to get to its intended destination is the function of the network layer. It is responsible for routing, which is the process of using a network layer address to determine the best path a packet will travel to its destination. Network layer protocols include Internet Protocol (IP, part of the TCP/IP protocol suite), and Internetwork Packet Exchange (IPX, part of the IPX/SPX protocol suite).

➤ *Data Link*—Divided into two sublayers (media access control [MAC] and logical link control [LLC]), the data-link layer is responsible for:

> ➤ Preparing data from upper layers to be transmitted over the physical medium by encapsulating upper-layer data into frames. This frame includes the source and destination of MAC addresses in the frame header.

> ➤ Converting data into bits, so it can be transmitted by the physical layer.

> ➤ Adding a cyclical redundancy check (CRC) to the end of a frame, which is used for error checking at the data-link layer.

Data-link protocols include High-Level Data Link Control (HDLC, the Cisco default encapsulation for serial connections), Synchronous Data Link Control, (SDLC, used in IBM networks), Link Access Procedure Balanced (LAPB, used with X.25), X.25 (a packet switching network), Serial Line Internet Protocol (SLIP, an older TCP/IP dial-up protocol), Point-to-Point (PPP, a newer dial-up protocol), Integrated Services Digital Network (ISDN, a dial-up digital service), and frame relay (a packet switching network).

➤ *Physical*—The lowest layer of the OSI model, the physical layer, defines the electrical functionality required to send and receive bits over a given physical medium. Specifications that define the voltage levels and physical components of a network are at the physical layer. Protocols are not specified at the physical layer, because they are implemented as software. Examples of the standards for sending data over the physical medium are Ethernet (the most widely used standard in networking), Token Ring (IBM's proprietary networking topology), and Fiber Distributed Data Interface (FDDI, a standard for fiber optic networks, commonly used as a backbone).

For more information, see Chapter 4 of *CCNA Routing and Switching Exam Cram, Second Edition.*

Question 20

The correct answer is c. Integrated Services Digital Network (ISDN) is a communications standard that uses digital telephone lines to transmit voice, data, and video. This is a dial-up service that is used on demand. ISDN protocol standards are defined by the International Telecommunication Union (ITU). Protocols that begin with the letter *E* (such as E.164) define ISDN standards on the existing telephone network. Protocols that begin with the letter *Q* (such as Q.931 and Q.921) specify ISDN switching and signaling standards, and protocols that begin with the letter *I* (such as I.430 and I.431) specify ISDN concepts, interfaces, and terminology. Answer a is incorrect, because protocols that begin with the letter *Q* (such as Q.931 and Q.921) specify ISDN switching and signaling standards. Answer b is incorrect, because protocols that begin with the letter *I* (such as I.430 and I.431) specify ISDN concepts, interfaces, and terminology. Answer d is incorrect, because protocols that begin with the letter *E* (such as E.164) define ISDN standards on the existing telephone network.

The bandwidth provided by ISDN can be divided into two categories: Primary Rate ISDN (PRI) and Basic Rate ISDN (BRI). Basic Rate ISDN (also known as 2B+1D) is composed of two B channels that are 64kbps each and one that is a 16kbps D channel. Basic Rate ISDN uses the two B channels to transmit data and uses the 16kbps D channel for control and signaling purposes. Primary Rate ISDN (also known as 23B+1D) is composed of 23 B channels that are 64kbps each and one that is a 64kbps D channel.

For more information, see Chapter 11 of *CCNA Routing and Switching Exam Cram, Second Edition.*

Question 21

The correct answer is d. The presentation layer functions as a translator and is responsible for text and data syntax translation, such as EBCDIC used in IBM systems and ASCII used in PCs and most computer systems. Presentation layer protocols include PICT, TIFF, JPEG, MIDI, MPEG, and Quick Time (audio/video application).

Answer a is incorrect, because SMTP, FTP, and EDI are all application layer protocols. Answer b is incorrect, because FTP and Telnet are application layer protocols. Answer c is incorrect, because Telnet is an application layer protocol. The application layer of the OSI model functions as the "window" for applications to access network resources and is responsible for verifying the identity of the destination node. Application layer protocols include Telnet, FTP, SMTP, WWW, EDI, and WAIS.

For more information, see Chapter 3 of *CCNA Routing and Switching Exam Cram, Second Edition.*

Question 22

The correct answer is b. ISDN is a communications standard that uses digital telephone lines to transmit voice, data, and video. ISDN is a dial-up service that is used on demand. The bandwidth provided by ISDN can be divided into two categories: Primary Rate ISDN (PRI) and Basic Rate ISDN (BRI). Basic Rate ISDN (also known as 2B+1D) is composed of two B channels that are 64kbps each and one that is a 16kbps D channel. Basic Rate ISDN uses the two B channels to transmit data and the 16kbps D channel for control and signaling purposes. Primary Rate ISDN (also known as 23B+1D) is composed of 23 B channels that are 64kbps each and one 64kbps D channel. Answer a is incorrect because the D channel used in Primary ISDN is 64kbps, not 16kbps. Answer c is incorrect, because Primary ISDN uses 23 B channels, not 23 D channels. Answer d is incorrect, because Primary ISDN uses one D channel, not two.

For more information, see Chapter 12 of *CCNA Routing and Switching Exam Cram, Second Edition.*

Question 23

The correct answer is b. RAM is used as the router's main working area and contains the running configuration. Answer a is incorrect, because the Flash is an erasable, programmable memory area that contains the Cisco IOS. Answer c is incorrect, because ROM is a physical chip installed on a router's motherboard that contains the bootstrap program. Answer d is incorrect, because NVRAM is used to *store* the startup or running configuration. The internal components of Cisco routers are comparable to that of conventional PCs. A Cisco router is similar to a PC in that it contains an operating system (known as the IOS), a bootstrap program (like the BIOS used in a PC), RAM (memory), and a processor (CPU). Unlike a PC, a Cisco router has no external drives to install software or to upgrade the IOS. Rather than use a floppy or CD drive to enter software as PCs do, Cisco routers are updated by downloading update information from a Trivial File Transfer Protocol (TFTP) server. The following list outlines the core components of Cisco routers:

➤ *ROM (Read-Only Memory)*—A physical chip installed on a router's motherboard that contains the bootstrap program, the Power On Self Test (POST), and the operating system software (Cisco IOS). When a Cisco router is first powered up, the bootstrap program and the POST are executed. ROM cannot be changed through software; the chip must be replaced if any modifications are needed.

➤ *RAM (Random Access Memory)*—As with PCs, RAM is used as the router's main working area and contains the running configuration. All information in RAM dissipates when the router is turned off.

➤ *NVRAM (Non-Volatile RAM)*—Used by Cisco routers to store the startup or running configuration. After a configuration is created, it exists and runs in RAM. Because RAM cannot permanently store this information, NVRAM is used to permanently store this configuration file. Information in NVRAM is preserved after the router is powered off.

➤ *Flash*—Essentially the programmable read-only memory (PROM) used in PCs, the Flash is an erasable, programmable memory area that contains the IOS. When a router's operating system needs to be up-graded, new software can be downloaded from a TFTP server to the router's Flash. Upgrading the IOS in this manner is typically more convenient than replacing the ROM chip in the router's motherboard. Information in Flash is retained when the router is powered off.

For more information, see Chapter 4 of *CCNA Routing and Switching Exam Cram, Second Edition.*

Question 24

The correct answer is b. The access list, **ACCESS-LIST 101 PERMIT TCP 195.20.200.0 0.0.0.255 151.120.0.0 0.0.255.255 EQ 21**, will allow all hosts on network **195.20.200.0** to have FTP access to network **151.120.0.0**. Answers a and c are incorrect, because the source and destination addresses are in the wrong order. The syntax for an extended IP access list specifies the source first and the destination second. Answer d is incorrect, because the port number used is 23, which specifies Telnet traffic, not FTP traffic. FTP uses port 21. The syntax for creating an extended IP access list is as follows:

```
ACCESS-LIST [access-list-number] [permit|deny] [protocol] [source]
  [wildcard mask] [destination] [wildmask] [operator][port]
```

➤ *ACCESS-LIST*—This is the command to denote an access list.

➤ *[access-list-number]*—This can be a number between 100 and 199 and is used to denote an extended IP access list.

➤ *[permit|deny]*—This will allow (permit) or disallow (deny) traffic specified in this access list.

➤ *[protocol]*—This is used to specify the protocol that will be filtered in this access list. Common values are TCP, UDP, and IP. (Use of IP will denote all IP protocols.)

➤ *[source]*—This specifies the source IP addressing information.

➤ *[wildcard mask]*—This can be optionally applied to further define the source. A wildmask can be used to control access to an entire IP network ID, rather than a single IP address. A wildcard mask will use the number 255 to mean "any" and the number 0 to mean "must match exactly." The terms *host* and *any* may also be used here to more quickly specify a single host or an entire network. The **HOST** keyword is the same as the wildcard mask 0.0.0.0. The **ANY** keyword is the same as the wildcard mask 255.255.255.255.

➤ *[destination]*—This specifies the destination IP addressing information.

➤ *[wildmask]*—This can be optionally applied to further define the destination IP addressing.

➤ *[operator]*—This can be optionally applied to define how to interpret the value entered in the **[port]** section of the access list. Common values are equal to the port specified (**EQ**), less than the port number specified (**LT**), and greater than the port number specified (**GT**).

➤ *[port]*—This specifies the port number this access list will act on. Common port numbers include 21 (FTP), 23 (TELNET), 25 (SMTP), 53 (DNS), 69 (TFTP), and 80 (HTTP).

For more information, see Chapter 13 of *CCNA Routing and Switching Exam Cram, Second Edition*.

Question 25

The correct answer is c. The three LMI signaling formats are: CISCO, ANSI, and Q933A. The Cisco default LMI type is CISCO. Based on this information, choices a, b, and d are incorrect.

Frame relay is a packet switching WAN technology that utilizes logical or virtual circuits to form a connection. Frame relay networks utilize Local Management Interface (LMI) signaling types, which are used to manage a frame relay connection. Cisco routers support three LMI signaling types: CISCO, ANSI, and Q933A. The Cisco default LMI type is CISCO. Frame relay is a packet switching technology that utilizes logical circuits to form a connection. This logical connection is identified by the data-link connection identifier

(DLCI). The DLCI is a unique identifier used to identify a specific frame relay connection. A frame relay connection is managed through a signaling format known as LMI. The LMI interface provides information about the DLCI values, as well as other connection-related information. Your Cisco router must utilize the same LMI signaling format as your service provider.

For more information, see Chapter 11 of *CCNA Routing and Switching Exam Cram, Second Edition.*

Question 26

The correct answer is d. The **ROUTER RIP/NETWORK 200.200.200.0** command will enable the Routing Information Protocol (RIP) on the router and have it advertise network 200.200.200.0. Answers a and b are incorrect, because **ROUTER RIP 200.200.200.0** and **NETWORK RIP** are not valid Cisco IOS commands. Answer c is incorrect, because the **ROUTER RIP** command does not have a hyphen between **ROUTER** and **RIP**. RIP is a distance vector routing protocol used to automatically update entries in a routing table. RIP routers update their routing tables by broadcasting the contents of their routing table every 30 seconds. To enable RIP on a Cisco router, the **ROUTER RIP** command is used. The syntax for the **ROUTER RIP** command is as follows:

```
ROUTER RIP
NETWORK [network id]
```

The **ROUTER RIP** command enables RIP on the router. The **NETWORK** command specifies the network ID of the route that will be advertised by the router.

For more information, see Chapter 11 of *CCNA Routing and Switching Exam Cram, Second Edition.*

Question 27

The correct answer is b. A Cisco router's startup sequence is divided into three basic operations. The POST executes, the operating system is loaded, and, finally, the startup configuration is loaded into RAM. The following steps detail the startup sequence:

1. The bootstrap program in ROM executes the POST, which performs a basic hardware-level check.

2. The Cisco IOS is loaded into memory per instructions from the boot system command. The IOS can be loaded from Flash, ROM, or a network location (TFTP server).

3. The router's configuration file is loaded into memory from NVRAM. If no configuration file is found, the setup program automatically initiates the dialog necessary to create a configuration file.

For more information, see Chapter 4 of *CCNA Routing and Switching Exam Cram, Second Edition.*

Question 28

The correct answer is c. The **CLEAR ACCESS-LIST COUNTERS 155** command will clear the previous matches from access list 155. The syntax for the **CLEAR ACCESS-LIST COUNTERS** command is: **CLEAR ACCESS-LIST COUNTERS [access list number]**.

Answer a is incorrect, because in the **CLEAR ACCESS-LIST COUNTERS** command, no hyphen is between **LIST** and **COUNTERS**. Answer b is incorrect, because **MATCHES** is not a valid Cisco IOS command. Answer d is incorrect, because in the **CLEAR ACCESS-LIST COUNTERS** command, a hyphen is between **ACCESS** and **LIST**.

For more information, see Chapter 13 of *CCNA Routing and Switching Exam Cram, Second Edition.*

Question 29

The correct answer is a. **ACCESS-LIST 800 DENY 20 40** and **ACCESS-LIST 800 PERMIT −1 −1** are the correct access lists to use in this situation. Answer b is incorrect, because the source and destination networks are in the wrong order. The syntax for a standard IPX access list specifies the source first, then the destination. Answer c is incorrect, because the **ACCESS-LIST** command uses a hyphen between **ACCESS** and **LIST**. Answer d is incorrect, because the syntax for an IPX access list does not begin with "IPX".

Access lists can be used to filter the traffic that is handled by a Cisco router. An IPX access list will filter IPX traffic and can be created as a standard IPX access list or an extended IPX access list. A standard IPX access list can be used to permit or deny access based on the source and destination IPX address information. The syntax for creating a standard IPX access list is:

```
ACCESS-LIST [access-list-number] [permit|deny] source
[source-node-mask] [destination] [[destination-node-mask]
```

➤ *ACCESS-LIST*—This command is followed by the **access list number.** When using a standard IPX access list, this can be a number between 800 and 899.

➤ *[permit\deny]*—This statement will either allow or disallow traffic in this access list.

➤ *source*—This statement is used to specify the source IPX addressing information. A "–1" can be used to signify "all networks."

➤ *[destination]*—This statement is used to specify destination addressing information. ("–1" may be used here as well.)

➤ *[source-node-mask] [destination-node-mask]*—These optional parameters control access to a portion of a network, similar to the wildmask that is used in IP access lists.

For more information, see Chapter 13 of *CCNA Routing and Switching Exam Cram, Second Edition.*

Question 30

The correct answer is b. **COPY TFTP STARTUP-CONFIG** will copy a configuration file from a TFTP server to the startup configuration in NVRAM. The startup configuration is a configuration file in a Cisco router that contains the commands used to set router-specific parameters. The startup configuration is stored in NVRAM (Non-Volatile RAM), and is loaded into RAM when the router starts up. Answer a is incorrect, because the correct syntax for the **COPY TFTP STARTUP-CONFIG** command has a hyphen between **STARTUP** and **CONFIG**. Answers c and d are incorrect, because the **COPY STARTUP CONFIG TFTP** command will copy the startup configuration *to* a TFTP server, not from one.

The running configuration is the startup configuration that is loaded into RAM and is the configuration used by the router when it is running. Because any information in RAM is lost when the router is powered off, the running configuration can be saved to the startup configuration, which is located in NVRAM, and will, therefore, be saved when the router is powered off. In addition, both the running and startup configuration files can be copied to and from a TFTP server, for backup purposes. The following list outlines the common commands used when working with configuration files:

➤ *COPY STARTUP-CONFIG RUNNING-CONFIG*—This will copy the startup configuration file in NVRAM to the running configuration in RAM.

➤ *COPY STARTUP-CONFIG TFTP*—This will copy the startup configuration file in NVRAM to a remote TFTP server.

➤ *COPY RUNNING-CONFIG STARTUP-CONFIG*—This will copy the running configuration from RAM to the startup configuration file in NVRAM.

➤ *COPY RUNNING-CONFIG TFTP*—This will copy the running configuration from RAM to a remote TFTP server.

➤ *COPY TFTP RUNNING-CONFIG*—This will copy a configuration file from a TFTP server to the router's running configuration in RAM.

➤ *COPY TFTP STARTUP-CONFIG*—This will copy a configuration file from a TFTP server to the startup configuration in NVRAM.

Note that the basic syntax of the **COPY** command specifies the source first, then the destination.

To modify a router's running configuration, you must be in global configuration mode. The **CONFIG T** command is used in privileged mode to access global configuration mode. Global configuration mode is displayed at the router prompt with "**config**" in parentheses (**RouterA(config)#**). To exit global configuration mode, use the keys Control and Z (Ctrl+Z).

For more information, see Chapter 8 of *CCNA Routing and Switching Exam Cram, Second Edition.*

Question 31

The correct answer is b. The **SHOW IP INTERFACE S0** command will display a list of all the IP access lists that are currently in use on interface **S0** of your Cisco router. Answer a is incorrect, because the **SHOW IP ACCESS-LISTS** command cannot be followed with an interface number. Answer c is incorrect, because **SHOW IP ACCESS-LISTS** will not display what interface an access list has been applied to, like the **SHOW INTERFACE** commands. Answer d is incorrect, because **SHOW IP INTERFACE-ACCESS-LISTS S0** is not a valid Cisco IOS command. Access lists can be used to filter the traffic that is handled by a Cisco router. Several Cisco IOS commands can be used to view the various access lists in use on a Cisco router. The following list details the most common Cisco IOS commands used to view access lists:

➤ *SHOW ACCESS-LISTS*—This will display all of the access lists in use on the router. It will also show each line of the access list and return the number of times a packet matched that line. This command can also be used with a specific access list number to display this detail on a single IP access list.

➤ *SHOW ACCESS-LISTS*—This will not display what interface an access list has been applied to, like the **SHOW INTERFACE** commands.

➤ *SHOW IP ACCESS-LIST*—This will display all IP access lists in use on the router. It will also show each line of the access list and return the number of times a packet matched that line. This command can also be used with a specific IP access list number to display this detail on a single IP access list. **SHOW IP ACCESS-LISTS** will not display what interface an access list has been applied to, like the **SHOW INTER-FACE** commands do.

➤ *SHOW IPX ACCESS-LIST*—This will display all IP access lists in use on the router. It will also show each line of the access list and return the number of times a packet matched that line. This command can also be used with a specific IPX access list number to display this detail on a single IPX access list. **SHOW IPX ACCESS-LISTS** will not display what interface an access list has been applied to, like the **SHOW INTERFACE** commands do.

➤ *SHOW IP INTERFACE [interface number]*—This will display the IP access lists that have been applied to a specific interface. It will not display any matches to the individual lines of the access list, like the **SHOW ACCESS-LIST** commands do.

➤ *SHOW IPX INTERFACE [interface number]*—This will display the IP access lists that have been applied to a specific interface. It will not display any matches to the individual lines of the access list, like the **SHOW ACCESS-LIST** commands do.

For more information, see Chapter 13 of *CCNA Routing and Switching Exam Cram, Second Edition.*

Question 32

The correct answer is c. Network layer protocols include IP (part of the TCP/IP protocol suite), and IPX (part of the IPX/SPX protocol suite). Answer a is incorrect, because SMTP and Telnet are application layer protocols. Answer b is incorrect, because TCP (part of the TCP/IP protocol suite), and SPX (part of the IPX/SPX protocol suite) are transport layer protocols. Answer d is incorrect, because SQL and RPC are session layer protocols.

For more information, see Chapter 4 of *CCNA Routing and Switching Exam Cram, Second Edition.*

Question 33

The correct answer is c. The OSI model defines a layered approach to network communication. Each layer of the OSI model adds its own layer-specific information and passes it on to the next layer, until it leaves the computer and goes out onto the network. This process is known as encapsulation, and it involves five basic steps:

1. User information is converted to data.

2. Data is converted to segments.

3. Segments are converted to packets or datagrams.

4. Packets and datagrams are converted to frames.

5. Frames are converted to bits.

For more information, see Chapter 4 of *CCNA Routing and Switching Exam Cram, Second Edition*.

Question 34

The correct answer is d. The OSI model defines standards used in networking and is composed of a seven-layer model. The transport layer of the OSI model is responsible for ensuring the reliable transfer of data by implementing connection-oriented services between nodes. Because no other layer of the OSI model performs this function, all of the other answers are incorrect.

The following list outlines the seven layers of the OSI model and their primary functions:

➤ *Application*—The "window" to networking used by programs; the application layer is responsible for:

➤ Verifying that the appropriate resources are present to initiate a connection with the destination node.

➤ Verifying the identity of the destination node.

Application layer protocols include Telnet, File Transfer Protocol (FTP), Simple Mail Transfer Protocol (SMTP), World Wide Web (WWW), Electronic Data Interchange (EDI), and Wide Area Information Server (WAIS).

➤ *Presentation*—Essentially a translator, the presentation layer is responsible for:

➤ Translating text and data syntax, such as Extended Binary-Coded Decimal Interchange Code (EBCDIC, used in IBM systems) and

American Standard Code for Information Interchange (ASCII, used in PC and most computer systems).

➤ Using Abstract Syntax Notation to perform data translation.

Presentation layer protocols include PICTure (PICT, Apple computer picture format), Tagged Image File Format (TIFF), Joint Photographic Experts Group (JPEG), Musical Instrument Digital Interface (MIDI), Motion Picture Expert Group (MPEG), and Quick Time (audio/video application).

➤ *Session*—Used to coordinate communication between nodes, the session layer is responsible for:

➤ Creating, maintaining, and ending a communication session.

➤ Coordinating service requests and responses that occur between nodes across the network.

Session layer protocols include Network File System (NFS, used by SUN Microsystems and Unix with TCP/IP), structured query language (SQL, used to define database information requests), remote procedure calls (RPC, used in Microsoft network communication), X Window (used by Unix terminals), AppleTalk Session Protocol (ASP, used by Apple computers), and Digital Network Architecture Session Control Protocol (DNA SCP, used by IBM).

➤ *Transport*—Reliable end-to-end communication is the primary function of the transport layer. Its many responsibilities include:

➤ Ensuring flow control so the amount of data being sent from one node will not overwhelm the destination node.

➤ Ensuring the ability of multiple applications to utilize a single transport (multiplcxing).

➤ Ensuring the reliable transfer of data. The transport layer implements connection-oriented services between nodes, which utilize a three-way handshake (synchronization, acknowledgment, and data transfer) to efficiently transfer data.

➤ Ensuring positive acknowledgment or the process of a node waiting for an acknowledgment from the destination node prior to sending data.

➤ Windowing, a form of flow control, that specifies how much data will be transferred between acknowledgments.

Transport layer protocols include Transmission Control Protocol (TCP, part of the TCP/IP protocol suite), and Sequenced Packet Exchange (SPX, part of the IPX/SPX protocol suite).

➤ *Network*—Selecting the appropriate path a packet should take to get to its intended destination is the function of the network layer. It is responsible for routing, which is the process of using a network layer address to determine the best path a packet will travel to its destination. Network layer protocols include Internet Protocol (IP, part of the TCP/IP protocol suite), and Internetwork Packet Exchange (IPX, part of the IPX/SPX protocol suite).

➤ *Data Link*—Divided into two sublayers (media access control [MAC] and logical link control [LLC]), the data-link layer is responsible for:

➤ Preparing data from upper layers to be transmitted over the physical medium by encapsulating upper-layer data into frames. This frame includes the source and destination of MAC addresses in the frame header.

➤ Converting data into bits, so it can be transmitted by the physical layer.

➤ Adding a cyclical redundancy check (CRC) to the end of a frame, which is used for error checking at the data-link layer.

Data-link protocols include High-Level Data Link Control (HDLC, the Cisco default encapsulation for serial connections), Synchronous Data Link Control, (SDLC, used in IBM networks), Link Access Procedure Balanced (LAPB, used with X.25), X.25 (a packet switching network), Serial Line Internet Protocol (SLIP, an older TCP/IP dial-up protocol), Point-to-Point (PPP, a newer dial-up protocol), Integrated Services Digital Network (ISDN, a dial-up digital service), and frame relay (a packet switching network).

➤ *Physical*—The lowest layer of the OSI model, the physical layer, defines the electrical functionality required to send and receive bits over a given physical medium. Specifications that define the voltage levels and physical components of a network are at the physical layer. Protocols are not specified at the physical layer, because they are implemented as software. Examples of the standards for sending data over the physical medium are Ethernet (the most widely used standard in networking), Token Ring (IBM's proprietary networking topology), and Fiber Distributed Data Interface (FDDI, a standard for fiber optic networks, commonly used as a backbone).

For more information, see Chapter 4 of *CCNA Routing and Switching Exam Cram, Second Edition.*

Question 35

The correct answer is b. Typing **TERMINAL NO EDITING** will disable the command history feature. It is enabled by default. All of the other answers, a, c, and d, are incorrect, because they are not valid Cisco IOS commands. The Cisco IOS includes a command history feature that can be used to store previously entered commands in a history buffer. Commands in the history buffer can be recalled by an operator at the router prompt, using Ctrl+P and Ctrl+N, saving configuration time. The command history feature is enabled by default, but it can be disabled, and the size of the history buffer can be modified. The following list details the common commands used when modifying the command history:

➤ *SHOW HISTORY*—Typing this command will display all of the commands currently stored in the history buffer. Pressing Ctrl+P will recall previous commands in the command history at the router prompt. (The up arrow will do this as well.) Pressing Ctrl+N will return to more recent commands in the command history after recalling previous commands. (The down arrow will do this as well.)

➤ *TERMINAL HISTORY SIZE (0–256)*—Typing this command will allow you to set the number of command lines the command history buffer will hold. You must enter a number between 0 and 256. (The default is 10 commands.)

➤ *TERMINAL NO EDITING*—Typing this command will disable the command history feature. It is enabled by default.

➤ *TERMINAL EDITING*—Typing this command will enable the command history feature if it has been disabled.

For more information, see Chapter 8 of *CCNA Routing and Switching Exam Cram, Second Edition*.

Question 36

The correct answer is c. **ACCESS-LIST 900 DENY −1 44 0 41 0** and **ACCESS-LIST 900 PERMIT −1 −1 0 −1 0** are the correct access lists to use in this situation. Answer a is incorrect, because no hyphen is between **ACCESS** and **LIST**. Answer b is incorrect, because the source and destination networks are in the wrong order. The syntax for an extended IPX access list specifies the source first, then the destination. Answer d is incorrect, because an extended IPX access list uses numbers between 900 and 999.

Access lists can be used to filter the traffic that is handled by a Cisco router. An IPX access list will filter IPX traffic and can be created as a standard IPX or an extended IPX access list. A standard IPX access list can be used to permit or deny access based on the source and destination IPX address information. An extended IPX access list can be used to filter traffic based the on source and destination IPX address information, as well as IPX protocol and IPX socket number. The syntax for creating an extended IPX access list is:

```
ACCESS-LIST [access-list-number] [permit|deny]  [protocol][source]
 [source-node-mask]  [source socket] [destination]
 [destination-node-mask] [destination socket]
```

➤ *ACCESS-LIST*—This command is followed by the **access list number**. When using a standard IPX access list, this can be a number between 800 and 899.

➤ *[permit|deny]*—These statements will either allow or disallow traffic in this access list.

➤ *[source]*—This statement is used to specify the source IPX addressing information. A "–1" can be used to signify "all networks."

➤ *[destination]*—This statement is used to specify destination addressing information. ("–1" may be used here as well.)

➤ *[source-node-mask] [destination-node-mask]*—These optional parameters control access to a portion of a network, similar to the wildmask used in IP access lists.

➤ *[protocol]*—This statement is used to specify the IPX protocol to filter. "–1" can be used to specify "all protocols."

➤ *[source socket]*—This statement is used to specify the IPX socket number to filter; "0" can be used to specify "all sockets."

For more information, see Chapter 13 of *CCNA Routing and Switching Exam Cram, Second Edition.*

Question 37

The correct answers are a, c, and g. Interior Gateway Routing Protocol (IGRP) is a distance vector protocol that is proprietary to Cisco and was designed to address the shortcomings of RIP. It has a maximum hop count of 255 and can make more informed decisions when selecting routes. IGRP considers hop count, delay, bandwidth, and reliability when determining the best path, making it a more efficient routing protocol than RIP.

RIP is a dynamic routing protocol used to automatically update entries in a routing table and is a Distance Vector Multicast Routing Protocol (DVMRP). RIP routers will broadcast the contents of their routing table every 30 seconds. This broadcast traffic does not make RIP very efficient for large networks. To select the best route, a RIP router selects the path with the fewest number of hops. (One hop is the distance from one router to another.) The maximum hop count used in RIP is 15. Any destination that requires more than 15 hops is considered unreachable by RIP. Answer b is incorrect, because RIP has a maximum hop count of 15. Answer d is incorrect, because RIP is not proprietary to Cisco. Answer e is incorrect, because RIP is not a link state protocol, nor is it proprietary to Cisco. Answer f is incorrect, because IGRP and RIP are distance vector protocols, not link state protocols.

For more information, see Chapter 11 of *CCNA Routing and Switching Exam Cram, Second Edition.*

Question 38

The correct answer is a. Flow control is a method used by TCP at the transport layer that ensures that the amount of data being sent from one node will not overwhelm the destination node. Three basic techniques are used in flow control:

➤ *Windowing*—This is the process of predetermining the amount of data that will be sent before an acknowledgment is expected.

➤ *Buffering*—This is a method of temporarily storing packets in memory buffers until they can be processed by the computer.

➤ *Source-Quench Messages*—These are messages sent by a receiving computer when its memory buffers are nearing capacity. A source-quench message is a device's way of telling the sending computer to slow down the rate of transmission.

Answers b and c are incorrect, because neither Buffering nor Source-Quench messages predetermine the amount of data that is to be sent. Answer d is incorrect, because multiplexing is the ability of multiple applications to utilize a single transport and is not used as part of flow control.

For more information, see Chapter 9 of *CCNA Routing and Switching Exam Cram, Second Edition.*

Question 39

The correct answer is b. Privileged mode contains the **CONFIGURE** command, which is used to access other configuration modes, such as global configuration and interface modes. Answers a and c are incorrect, because user mode does not contain the **CONFIGURE** command. Answer d is incorrect, because no administrative mode is on a Cisco router.

The Cisco IOS uses a command interpreter known as EXEC, which contains two primary modes: user mode and privileged mode. User mode is used to display basic router system information and to connect to remote devices. The commands in user mode are limited in capability. When you first log into a router, you are placed in user mode, where the router prompt is followed by an angle bracket (**ROUTER>**). Privileged mode is used to modify and view the configuration of the router, as well as set IOS parameters. In privileged mode, the router prompt is followed by a pound sign (**ROUTER#**). To enter privileged mode, enter the **ENABLE** command from user mode.

For more information, see Chapter 8 of *CCNA Routing and Switching Exam Cram, Second Edition*.

Question 40

The correct answer is a. Typing a single question mark at the router prompt will list all available commands for that particular router mode. All of the other choices are invalid Cisco ISO commands. The Cisco IOS includes a context-sensitive help feature that can assist an operator when entering commands into a Cisco router. Context-sensitive help provides information on available IOS commands, command line syntax, and available command keywords. Accessing the features of context-sensitive help involves the correct use of the question mark key (?). The following details the common use of the question mark (?) key in context-sensitive help:

➤ *ROUTER#?*—Typing a single question mark at the router prompt will list all available commands for that particular router mode.

➤ *ROUTER#<COMMAND> ?*—Typing a command followed by a space, then a question mark, will display any keywords that can be used with that command. For example, typing **CLOCK ?** at the router prompt will return **SET**, the keyword used with the **CLOCK** command.

➤ *ROUTER#<PARTIAL COMMAND>?*—Typing a partial command followed *immediately* by a question mark (no space between partial command and question mark) will display a list of commands that begins with the partial command entry. For example, typing **CL?** at the router prompt will return **CLEAR** and **CLOCK**, the two IOS commands that begin with "CL". Note that if a space is included between a partial or complete command and a question mark, help will interpret this as a keyword request, and it could result in an "ambiguous command" message. For example, typing **CL** ? (CL, a space, then a question mark) will result in the message "Ambiguous command: "cl " ", because two commands begin with "CL:" (**CLOCK** and **CLEAR**).

For more information, see Chapter 8 of *CCNA Routing and Switching Exam Cram, Second Edition.*

Question 41

The correct answer is g. All of the answers are valid entries for the port field in an extended IP access list except for IP. Therefore, answers a, b, c, d, e, and f are incorrect. IP is used in the protocol field of an extended IP access list. When IP is used as a protocol, it is interpreted by the router as meaning "all IP protocols, and no port number can be used". The syntax for creating an extended IP access list is as follows:

```
ACCESS-LIST [access-list-number] [permit|deny] [protocol]
[source] [wildcard mask]  [destination] [wildmask] [operator]
[port]
```

➤ *ACCESS-LIST*—This is the command to denote an access list.

➤ *[access-list-number]*—This can be a number between 100 and 199. It is used to denote an extended IP access list.

➤ *[permit|deny]*—This will allow (permit) or disallow (deny) traffic specified in this access list.

➤ *[protocol]*—This is used to specify the protocol that will be filtered in this access list. Common values are TCP, UDP, and IP. (Use of IP will denote all IP protocols.)

➤ *[source]*—This specifies the source IP addressing information.

➤ *[wildcard mask]*—This can be optionally applied to further define the source. A wildmask can be used to control access to an entire IP network ID rather than a single IP address. A wildcard mask will use the number 255 to mean "any" and the number 0 to mean "must match exactly." The terms **HOST** and **ANY** may also be used here to more quickly specify a single host or an entire network. The **HOST** keyword is the same as the wildcard mask 0.0.0.0. The **ANY** keyword is the same as the wildcard mask 255.255.255.255.

➤ *[destination]*—This specifies the destination IP addressing information.

➤ *[wildmask]*—This can be optionally applied to further define the destination IP addressing.

➤ *[operator]*—This can be optionally applied to define how to interpret the value entered in the **[port]** section of the access list. Common values are EQ (equal to the port specified), LT (less than the port number specified), and GT (greater than the port number specified).

➤ *[port]*—This specifies the port number this access list will act on. Common port numbers include 21 (FTP), 23 (Telnet), 25 (SMTP), 53 (DNS), 69 (TFTP), and 80 (HTTP).

For more information, see Chapter 13 of *CCNA Routing and Switching Exam Cram, Second Edition*.

Question 42

The correct answer is d. The access list, **ACCESS-LIST 1000 DENY 8E.1234.1234.1234 4**, will deny the advertisement of file service from Netware server **8E.1234.1234.1234 4**. Answer a is incorrect, because in the **ACCESS-LIST** command, a hyphen is between **ACCESS** and **LIST**. Answer b is incorrect, because an IPX Service Advertising Protocol (SAP) access list uses numbers between 1000 and 1099. Answer c is incorrect, because the file service number is 7, which denotes a Netware printer server.

An IPX SAP filter can be applied to a router's interface to filter the SAP advertisement traffic passed by the router. Novell Netware servers will broadcast a list of their available resources every 60 seconds. This is known as a SAP advertisement. Cisco routers will record this information into their own SAP table for advertisement to other segments. By filtering the amount of SAP traffic with an IPX SAP access list, network traffic can be controlled and/or reduced. The syntax for a SAP access list is as follows:

```
ACCESS-LIST [access-list-number] [permit|deny]    [source network]
 [service type]
```

➤ *ACCESS-LIST*—This command is followed by the **access list number**. When using an IPX SAP access list, this can be a number between 1000 and 1099.

➤ *[permit|deny]*—This statement will either allow or disallow traffic in this access list.

➤ *[source network]*—This statement is used to specify the source IPX network. A "–1" can be used to signify "all networks".

➤ *[service type]*—This statement is used to specify the type of service to be filtered. Common IPX service types include a Netware file server, specified as number 4, and a Netware printer server, specified as number 7.

For more information, see Chapter 13 of *CCNA Routing and Switching Exam Cram, Second Edition.*

Question 43

The correct answer is d. The DEMARC is the point at the customer's premises where CPE devices connect and is usually a large punch-down board located in a wiring closet. Answer a is incorrect, because **Customer Premises Equipment** are the communication devices that exist at the customer's location. Answer b is incorrect, because the CO, or Central Office, is the telephone company's office that acts as the central communication point for the customer. Answer c is incorrect, because the local loop is the cabling that extends from the DEMARC to the telephone company's central office (CO).

The most common form of WAN connectivity utilizes telecommunication technology. When WANs communicate over telecommunication lines, several components are used to facilitate the connection. The basic process of WAN communication begins with a call placed using Customer Premises Equipment (CPE). The CPE connects to the DEMARC, which passes the data through the local loop to the central office, which acts as the central distribution point for sending and receiving information. The following list details the common components of WAN communication:

➤ *Customer Premises Equipment*—These are the communication devices (telephones, modems, and so forth) that exist at the customer's location.

➤ *DEMARC*—This is short for demarcation; the DEMARC is the point at the customer's premises where CPE devices connect. The telephone company owns the DEMARC, which is usually a large punch-down board located in a wiring closet.

➤ *Local Loop*—This is the cabling that extends from the DEMARC to the telephone company's central office (CO).

➤ *Central Office*—This is the telephone company's office that acts as the central communication point for the customer.

For more information, see Chapter 11 of *CCNA Routing and Switching Exam Cram, Second Edition.*

Question 44

The correct answers are a and d. A switch is a data-link layer connectivity device used to segment a network. A switch resides at the data-link layer of the OSI model and can filter traffic based on the data-link address of a packet or a MAC address. (The MAC address is the address burned into the network adapter card by the manufacturer.)

Cisco switches employ two basic methods of forwarding packets: cut-through and store-and-forward. Store-and-forward switches will copy an incoming packet to the local buffer and perform error checking on the packet before sending it on to its destination. If the packet contains an error, it is discarded. Cut-through switches will read the destination address of an incoming packet and immediately search its switch table for a destination port. Cut-through switches will not perform error checking of the packet. The fact that cut-through switches do not perform error checking or copy the packet to the buffer before processing means they experience less latency (delay) than store-and-forward switches. Answer b is incorrect, because cut-through switches do not copy an incoming packet to their buffer. Answer c is incorrect, because store-and-forward switches experience more latency than cut-through switches, not less.

For more information, see Chapter 6 of *CCNA Routing and Switching Exam Cram, Second Edition.*

Question 45

The correct answer is a. DNA SCP (used by IBM) is a session layer protocol. All of the other protocols listed are data-link protocols, therefore, answers b, c, d, e, f, and g are incorrect. The data-link layer of the OSI model is responsible for preparing data from upper layers to be transmitted over the physical medium by encapsulating upper-layer data into frames. Data-link protocols include HDLC (the Cisco default encapsulation for serial connections), SDLC (used in IBM networks), LAPB (used with X.25), X.25 (a packet switching network), SLIP (an older TCP/IP dial-up protocol), PPP (a newer dial-up protocol), ISDN (a dial-up digital service), and frame relay (a packet switching network).

For more information, see Chapter 11 of *CCNA Routing and Switching Exam Cram, Second Edition.*

Question 46

The correct answer is d. RAM is used as the router's main working area and contains the running configuration. Answer a is incorrect, because ROM is the chip on a router that contains the POST and IOS image. Answer b is incorrect, because NVRAM is used by a Cisco router to store the startup configuration. Answer c is incorrect, because the Flash is an erasable, programmable memory area that contains the Cisco operating system software (IOS).

The internal components of Cisco routers are comparable to that of conventional PCs. A Cisco router is similar to a PC in that it contains an operating system (known as the IOS), a bootstrap program (like the BIOS used in a PC), RAM (memory), and a processor (CPU). Unlike a PC, a Cisco router has no external drives to install software or upgrade the IOS. Rather than use a floppy or CD drive to enter software as PCs do, Cisco routers are updated by downloading update information from a TFTP server. The following list outlines the core components of Cisco routers:

➤ *ROM (Read-Only Memory)*—This is a physical chip installed on a router's motherboard that contains the bootstrap program, the POST, and the operating system software (Cisco IOS). When a Cisco router is first powered up, the bootstrap program and the POST are executed. ROM cannot be changed through software; the chip must be replaced if any modifications are needed.

➤ *RAM (Random Access Memory)*—As with PCs, RAM is used as the router's main working area and contains the running configuration. All information in RAM dissipates when the router is turned off.

➤ *NVRAM (Non-Volatile RAM)*—This is used by Cisco routers to store the startup or running configuration. After a configuration is created, it exists and runs in RAM. Because RAM cannot permanently store this information, NVRAM is used to permanently store this configuration file. Information in NVRAM is preserved after the router is powered off.

➤ *Flash*—This is essentially the PROM used in PCs; the Flash is an erasable, programmable memory area that contains the Cisco operating system software (IOS). When a router's operating system needs to be upgraded, new software can be downloaded from a TFTP server to the router's Flash. Upgrading the IOS in this manner is typically more convenient than replacing the ROM chip in the router's motherboard. Information in Flash is retained when the router is powered off.

For more information, see Chapter 8 of *CCNA Routing and Switching Exam Cram, Second Edition*.

Question 47

The correct answer is a. The **BANNER MOTD # / Welcome to my router #** command will configure a login banner for this router. The other choices are invalid Cisco IOS commands, therefore answers b, c, and d are incorrect. A Cisco router can be configured to display a message when users log in. This message is known as a logon banner and can be configured using the **BANNER MOTD** command. The **BANNER MOTD** command must be entered in global configuration mode, which can be accessed from privileged mode using the **CONFIG T** command. The correct syntax for setting a logon banner is as follows:

```
BANNER MOTD #  (note that the pound sign (#)
 is a delimiting character of your choice)
[Message text]  #
```

An example of a logon banner would be:

```
Router(config)#banner motd #
Enter TEXT message.  End with the character
 '#'.
This is the message text of my logon banner #
Router(config)#^Z
```

For more information, see Chapter 8 of *CCNA Routing and Switching Exam Cram, Second Edition.*

Question 48

The correct answer is c. The **SHOW INTERFACE** command is used to display the configuration information of all interfaces. If the **SHOW INTERFACE** is issued with a specific port (that is, **SHOW INTERFACE E0**), then only the configuration of that interface will be displayed. The **SHOW INTERFACE** command can be issued from user or privileged mode. The other answers are invalid Cisco IOS commands, therefore answers a, b, and d are incorrect.

For more information, see Chapter 8 of *CCNA Routing and Switching Exam Cram, Second Edition.*

Question 49

The correct answer is d. The range of numbers used to specify an extended IPX access list is 900 to 999. Access lists can be easily identified by referring to the number after the **ACCESS-LIST** command. The five major types of access lists utilize the following ranges of numbers:

➤ *IP Standard Access List—1–99*

➤ *IP Extended Access List—100–199*

➤ *IPX Standard Access List—800–899*

➤ *IPX Extended Access List—900–999*

➤ *IPX SAP Access List—1000–1099*

For more information, see Chapter 13 of *CCNA Routing and Switching Exam Cram, Second Edition.*

Question 50

The correct answer is d. This access list is invalid because no hyphen is between **ACCESS** and **LIST**. Because this access list is invalid, answers a, b, and c are incorrect.

Access lists can be used to filter the traffic handled by a Cisco router. An IP access list will filter IP traffic and can be created as a standard IP access list or an extended IP access list. A standard IP access list can be used to permit or deny access based on IP addressing information only and can only act on the source IP addressing information. The syntax for creating a standard IP access list is:

```
ACCESS-LIST [access-list-number] [permit|deny]  source
[wildcard mask]
```

➤ *ACCESS-LIST*—This command is followed by the **access list number**. When using a standard IP access list, this can be a number between 1 and 99.

➤ *[permit|deny]*—This statement will either allow or disallow traffic in this access list.

➤ *source*—This statement is used to specify the source IP addressing information. A standard IP access list can only act on a source address.

➤ *[wildcard mask]*—This is an optional parameter in a standard IP access list. If no wildcard mask is specified, the access list will only act on a single IP address that is specified in the **source**. A wildmask can be used to control access to an entire IP network ID rather than a single IP address. A **wildcard mask** will use the number 255 to mean "any" and the number 0 to mean "must match exactly."

For more information, see Chapter 13 of *CCNA Routing and Switching Exam Cram, Second Edition.*

Question 51

The correct answer is a. The number, 9999.8888.7777, is the node ID of IPX address 7g.9999.8888.7777. Answers b and c are incorrect, because they both combine the node ID and the network ID. Answer d is incorrect, because 7g is the network ID of the address. The IPX/SPX protocol is used in Novell NetWare networks. An IPX address is a network layer, hierarchical address that is similar in format to an IP address in that IPX addresses contain a network ID and a node ID.

The network ID of an IPX address identifies the network that an IPX node is a part of, and will be the same for all IPX computers in the same network. The node ID of an IPX address is unique to each host on an IPX network and is assigned dynamically using the MAC address of the computer. This makes configuring IPX/SPX addresses much easier than TCP/IP addresses.

The network ID of an IPX address is assigned by an administrator and can be up to 32 bits in length. The network ID is usually expressed in decimal notation without the leading zeros. The node ID of an IPX address is 48 bits in length and can be expressed in decimal notation by four digits separated by three periods. A complete IPX address is expressed in decimal notation by combining the network ID with the node ID. For example, if an IPX host has a node ID of 1234.1234.1234 and is part of IPX network 2c, the IPX address for that host would be 2c.1234.1234.1234.

For more information, see Chapter 12 of *CCNA Routing and Switching Exam Cram, Second Edition.*

Question 52

The correct answers are a and c. A routing table can be of two basic types: static and dynamic. A static routing table is one that has its entries entered manually by an operator. On a Cisco router, a static route can be entered using the IP ROUTE command. A dynamic routing table is one that has its entries entered automatically by the router, through the use of a routing protocol, such as RIP or IGRP. Answer b is incorrect, because IP and IPX are routable protocols, not routing protocols. Answer d is incorrect, because UPDATE is not a valid Cisco IOS command.

For more information, see Chapter 11 of *CCNA Routing and Switching Exam Cram, Second Edition.*

Question 53

The correct answers are b, c, d, and e. Answer a is incorrect, because new horizon is not a valid method used to prevent routing loops. Answer f is incorrect, because horizon poisoning is not a valid method of preventing routing loops, which is a problem that can occur with distance vector routing.

A routing loop occurs when a link on a router fails, and some routers do not receive the update information. The routers that do not have the updated information in their routing table will advertise to other routers that they have a route to the failed link and will begin receiving the packets destined for that network. Because this route is failed and no real way exists to route the packet to its destination, packets will end up getting passed back and forth between routers and never reach their destination. To combat the problem of routing loops, distance vector routing protocols utilize four basic techniques: maximum hop mount, split horizon, route poisoning, and hold-downs. The following describes the function of each:

➤ *Maximum Hop Count*—Maximum hop count involves setting a maximum hop count that a packet can travel before it is discarded. RIP uses a maximum hop count of 15, so in the event of a routing loop, a packet that travels more than 15 hops will be discarded.

➤ *Split Horizon*—This method states that a packet cannot be sent back in the direction that it was received, thus preventing the routing loop of two routers from passing a packet back and forth between each other.

➤ *Route Poisoning*—When a router using a distance vector routing protocol has a link that fails, it will "poison" its route by setting the hop count of the failed link to a number over the maximum hop count, so other routers will deem the route unreachable. In the case of RIP, a router can poison a route by setting a hop count of 16. (Fifteen is the maximum hop count used by RIP.)

➤ *Hold-Downs*—A hold-down is a specified amount of time a router will wait before updating its routing table. Hold-downs prevent routing loops by preventing routers from updating new information too quickly, allowing time for a failed link to be reestablished.

For more information, see Chapter 11 of *CCNA Routing and Switching Exam Cram, Second Edition.*

Question 54

The correct answer is b. The IP addresses range from 192.23.98.49 to 192.23.98.62 and are valid host IDs for the nodes in this subnet. Based on this information, all of the other answers are incorrect. When subnetting IP addresses, host IDs are grouped together according to the incremental value of the subnet mask in use. The first address in this identifies the subnet (subnet ID) and is not used for a host ID. The last address in this range is used by the router as the broadcast address, and is also not used as a host ID. In this question, the subnet mask used is 240, which will divide subnets in groups of 16 (192.23.98.16 through 192.23.98.31, 192.23.98.32 through 192.23.98.47, 192.23.98.48 through 192.23.98.63, and so forth). The broadcast address is specified as 192.23.98.63, which places this address in subnet 192.23.98.48 through 192.23.98.63. The first address is used as the subnet ID and the last address is used for the broadcast address, making the valid range of host IDs 192.23.98.49 through 192.23.98.62. The following list details the incremental values of IP subnet masks:

➤ *Subnet mask 192*—Increments in 64

➤ *Subnet mask 224*—Increments in 32

➤ *Subnet mask 240*—Increments in 16

➤ *Subnet mask 248*—Increments in 8

➤ *Subnet mask 252*—Increments in 4

➤ *Subnet mask 254*—Increments in 2

➤ *Subnet mask 255*—Increments in 1

For more information, see Chapter 13 of *CCNA Routing and Switching Exam Cram, Second Edition.*

Question 55

The correct answer is a. **CONFIGURE TERMINAL** is used to enter configuration commands into the router from the console port or through Telnet. Answer b is incorrect, because the **CONFIGURE NETWORK (CONFIG NET)** command is used to copy the configuration file from a TFTP server into the router's RAM. Answer c is incorrect, because the **CONFIGURE OVERWRITE-NETWORK (CONFIG O)** is used to copy a configuration file into NVRAM from a TFTP server. Answer d is incorrect, because the **CONFIGURE MEMORY (CONFIG MEM)** command is used to execute the configuration stored in NVRAM and will copy the startup configuration to the running configuration.

The **CONFIGURE** command, executed in privileged mode, is used to enter the configuration mode of a Cisco router. The four parameters used with the **CONFIGURE** command are **TERMINAL, MEMORY, OVERWRITE-NETWORK,** and **NETWORK.** The following list outlines the function of these parameters:

➤ *CONFIGURE TERMINAL (CONFIG T)*—This command is used to enter configuration commands into the router from the console port or through Telnet.

➤ *CONFIGURE NETWORK (CONFIG NET)*—This command is used to copy the configuration file from a TFTP server into the router's RAM.

➤ *CONFIGURE OVERWRITE-NETWORK (CONFIG O)*—This command is used to copy a configuration file into NVRAM from a TFTP server. Use of the **CONFIGURE OVERWRITE-NETWORK** will not alter the running configuration.

➤ *CONFIGURE MEMORY (CONFIG MEM)*—This command is used to execute the configuration stored in NVRAM. It will copy the startup configuration to the running configuration.

For more information, see Chapter 8 of *CCNA Routing and Switching Exam Cram, Second Edition.*

Question 56

The correct answer is a. The **ENABLE PASSWORD Swordfish** command will set the enable password to "**Swordfish**" on a Cisco router; all other answers are invalid Cisco IOS commands. The enable password is used to control access to the privileged mode of the router. If an enable password is specified, it must be entered after the **ENABLE** command to successfully access privileged mode. To configure the enable password, use the **ENABLE PASSWORD** [*password*] command. Answers b and c are not valid Cisco IOS commands. Answer d is incorrect, because all passwords are case sensitive.

For more information, see Chapter 8 of *CCNA Routing and Switching Exam Cram, Second Edition.*

Question 57

The correct answer is b. Plain old telephone service (POTS) is a standard analog phone line and is used as a dial-up service through modems. Answer a is incorrect, because ISDN is a communications standard dial-up service that uses digital telephone lines. Answers c and d are incorrect, because frame relay and X.25 are packet switching technologies. Packet switching is a process of sending data over WAN links through the use of logical circuits, where data is encapsulated into packets and routed by the service provider through various switching points.

For more information, see Chapter 11 of *CCNA Routing and Switching Exam Cram, Second Edition.*

Question 58

The correct answer is b. The **ROUTER IGRP 11 NETWORK 223.20.200.0** command will enable IGRP on the router and have it advertise **network 223.20.200.0** in **autonomous system number 11**. Answer a is incorrect, because the **autonomous system number** is supposed to be entered after the **ROUTER IGRP** command. Answer c is incorrect, because the **ROUTER IGRP** command does not have a hyphen between **ROUTER** and **IGRP**. Answer d is incorrect, because the **ROUTER IGRP** command does not use **AS** after **ROUTER IGRP**.

IGRP is a distance vector protocol that is proprietary to Cisco and was designed to address the shortcomings of RIP. IGRP has a maximum hop count of 255 and can make more informed decisions when selecting routes. IGRP considers hop count, delay, bandwidth, and reliability when determining the best path, making it a more efficient routing protocol than RIP. To enable IGRP on a Cisco router, use the **ROUTER IGRP** command. The syntax for the **ROUTER IGRP** command is as follows:

```
ROUTER IGRP [autonomous system number]
NETWORK [network ID]
```

➤ *ROUTER IGRP*—This command enables IGRP on the router.

➤ *[autonomous system number]*—This identifies the group of IGRP routers that this router will communicate with. All routers must have the same **autonomous system number** to communicate.

➤ *NETWORK*—This command specifies the **network ID** that this router will advertise.

For more information, see Chapter 11 of *CCNA Routing and Switching Exam Cram, Second Edition.*

Question 59

The correct answer is b. RAM is used as the router's main working area and contains the running configuration. All information in RAM dissipates when the router is turned off. All of the other answers are capable of storing information after the router is powered off. Therefore, answers a, c, and d are incorrect. The following outlines the core components of Cisco routers:

➤ *ROM (Read-Only Memory)*—This is a physical chip installed on a router's motherboard that contains the bootstrap program, the POST, and the operating system software (Cisco IOS). When a Cisco router is first powered up, the bootstrap program and the POST are executed. ROM cannot be changed through software; the chip must be replaced if any modifications are needed.

➤ *RAM (Random Access Memory)*—As with PCs, RAM is used as the router's main working area and contains the running configuration. All information in RAM dissipates when the router is turned off.

➤ *NVRAM (Non-Volatile RAM)*—This is used by Cisco routers to store the startup or running configuration. After a configuration is created, it exists and runs in RAM. Because RAM cannot permanently store this information, NVRAM is used to permanently store this configuration file. Information in NVRAM is preserved after the router is powered off.

➤ *Flash*—Essentially, this is the PROM used in PCs. The Flash is an erasable, programmable memory area that contains the Cisco operating system software (IOS). When a router's operating system needs to be upgraded, new software can be downloaded from a TFTP server to the router's Flash. Upgrading the IOS in this manner is typically more convenient than replacing the ROM chip in the router's motherboard. Information in Flash is retained when the router is powered off.

For more information, see Chapter 8 of *CCNA Routing and Switching Exam Cram, Second Edition.*

Question 60

The correct answers are a and b. The Cisco IOS uses a command interpreter known as EXEC, which contains two primary modes: user mode and privileged mode. User mode is used to display basic router system information and to connect to remote devices. The commands in user mode are limited in capability. When you first log into a router, you are placed in user mode. In user mode, the router prompt is followed by an angle bracket (**ROUTER>**).

Privileged mode is used to modify and view the configuration of the router, as well as set IOS parameters. Privileged mode also contains the **CONFIGURE** command, which is used to access other configuration modes, such as global configuration mode and interface mode. In privileged mode, the router prompt is followed by a pound sign (**ROUTER#**). To enter privileged mode, enter the **ENABLE** command from user mode. Answers c and d are incorrect, because no executive or settings modes exist on a Cisco router.

For more information, see Chapter 8 of *CCNA Routing and Switching Exam Cram, Second Edition.*

Question 61

The correct answer is b. The router is in privileged mode, as evidenced by the router prompt being followed by a pound sign (#). Since privileged mode is the only mode that displays in this manner, all of the other answers are incorrect.

When you first log into a router, you are placed in user mode. The Cisco IOS uses a command interpreter known as **EXEC**, which contains two primary modes: user mode and privileged mode. User mode is used to display basic router system information and to connect to remote devices. The commands in user mode are limited in capability. In user mode, the router prompt is followed by an angle bracket (**ROUTER>**). Privileged mode is used to modify and view the configuration of the router, as well as set IOS parameters. Privileged mode also contains the **CONFIGURE** command, which is used to access other configuration modes, such as global configuration mode and interface mode. In privileged mode, the router prompt is followed by a pound sign (**ROUTER#**). To enter privileged mode, enter the **ENABLE** command from user mode.

For more information, see Chapter 8 of *CCNA Routing and Switching Exam Cram, Second Edition.*

Question 62

The correct answer is a. **CONFIGURE TERMINAL (CONFIG T)** is used to enter configuration commands into the router from the console port or through Telnet. Since the **CONFIGURE TERMINAL** command is the only Cisco IOS command that will perform this operation, all of the other answers are incorrect. The **CONFIGURE** command, executed in privileged mode, is used to enter the configuration mode of a Cisco router. The four parameters used with the **CONFIGURE** command are **TERMINAL, MEMORY, OVERWRITE-NETWORK,** and **NETWORK.**

The following list outlines the functions of these parameters:

➤ *CONFIGURE TERMINAL (CONFIG T)*—This command is used to enter configuration commands into the router from the console port or through Telnet.

➤ *CONFIGURE NETWORK (CONFIG NET)*—This command is used to copy the configuration file from a TFTP server into the router's RAM.

➤ *CONFIGURE OVERWRITE-NETWORK (CONFIG O)*—This command is used to copy a configuration file into NVRAM from a TFTP server. Use of the **CONFIGURE OVERWRITE-NETWORK** command will not alter the running configuration.

➤ *CONFIGURE MEMORY (CONFIG MEM)*—This command is used to execute the configuration stored in NVRAM. It will copy the startup configuration to the running configuration.

For more information, see Chapter 8 of *CCNA Routing and Switching Exam Cram, Second Edition.*

Question 63

The correct answer is a. The **IPX OUTPUT-SAP-FILTER 1080** is the correct command to use in this question to apply the access list **1080** to the outgoing interface of the router. Answer b is incorrect, because the command must begin with IPX. Answer c is incorrect, because the **IPX OUTPUT-SAP-FILTER** contains hyphens between **IPX**, **SAP**, and **FILTER**. Answer d is incorrect, because the **IPX ACCESS-GROUP** command is used to apply a standard to an extended IPX access list, not an IPX SAP access list.

For an IPX SAP access list to filter traffic, it must be applied to an interface. This is done using the **IPX-OUTPUT/INPUT-SAP-FILTER** command. This command must be entered in interface configuration mode. To access interface configuration mode, use the **INTERFACE [interface-number]** command from global configuration mode. When the router is in interface configuration mode, the router prompt displays "**config-if**" in parentheses, followed by a pound sign (**ROUTER(config-if)#**). The **IPX-OUTPUT/ INPUT-SAP-FILTER** command has the following syntax:

```
IPX[output|input]-SAP-FILTER  [access list number]
```

The **output|input** parameter specifies where the access list will be applied on the router. **OUTPUT** will cause the router to apply the access list to all outgoing packets, and **INPUT** will cause the router to apply the access list to all incoming packets.

For more information, see Chapter 12 of *CCNA Routing and Switching Exam Cram, Second Edition.*

Question 64

The correct answer is d. IGRP is a distance vector routing protocol. Answer a is incorrect, because RIP is a distance vector routing protocol. Answer b is incorrect, because open shortest path first (OSPF) is a link state routing protocol. Answer c is incorrect, because (EIGRP) Enhanced Interior Gateway Routing Protocol is considered a balanced hybrid routing protocol that combines the features of link state and distance vector. Answers e and f are incorrect, because IP and IPX are routed protocols and are not used to update routing tables. A Cisco router cannot route anything unless an entry is in its routing table that contains information on where to send the packet.

A routing table can be of two basic types: static and dynamic. A static routing table is one that has its entries entered manually by an operator. On a Cisco router, a static route can be entered using the **IP ROUTE** command. A dynamic routing table is one that has its entries entered automatically by the router, through the use of a routing protocol. A dynamic routing protocol provides for automatic discovery of routes and eliminates the need for static routes.

Routing protocols generally fall into one of two categories: link state and distance vector. Distance vector routing protocols dynamically update their routing tables by broadcasting their own routing information at specified intervals. Distance vector routing protocols include RIP and IGRP. Link state routing protocols, such as OSPF, update their routing tables by sending multicast "hello" messages to their neighbors and only send update information when a change occurs in the network.

For more information, see Chapter 11 of *CCNA Routing and Switching Exam Cram, Second Edition.*

Question 65

The correct answer is b. The subnet mask, 255.255.224.0, uses a total of 19 bits. An IP address is a 32-bit network layer address that is composed of a network ID and a host ID. A subnet mask is used to distinguish the Network portion of an IP address from the Host portion. The 32 bits in an IP address are expressed in dotted decimal notation in four octets (four 8-bit numbers separated by periods, that is, 208.100.50.90). The subnet mask, 255.255.224.0, is said to use 19 bits, because it takes 19 of 32 bit positions to create. This address, 255.255.224.0, can be expressed in binary as 11111111.11111111.11100000.00000000, which is 2 octets using 8 bits each (16) and 3 bits used in the third octet to make the number 224. Because the subnet mask is using 19 bits, and an IP address is a total of 32 bits, 13 bits remain to create host IDs on this network. Because subnet mask 255.255.224.0 uses a total of 19 bits, all of the other answers are incorrect.

For more information, see Chapter 8 of *CCNA Routing and Switching Exam Cram, Second Edition.*

Question 66

The correct answer is c. The **HELP** command is used to provide basic instructions on how to use the context-sensitive help feature of Cisco routers. Answer a is incorrect, because typing **HELP** along with a command name is not a valid use for the **HELP** command in Cisco routers. Answer b is incorrect, because the **HELP** command is a valid Cisco IOS command. Answer d is incorrect, because the **HELP** command does not contain any graphical display. The Cisco IOS includes a context-sensitive help feature that can assist an operator when entering commands into a Cisco router. Context-sensitive help provides information on available IOS commands, command line syntax, and available command keywords. Accessing the features of context-sensitive help involves the correct use of the question mark key (?). The following list details the common use of the question mark (?) key in context-sensitive help:

➤ *ROUTER#?*—Typing a single question mark at the router prompt will list all available commands for that particular router mode.

➤ *ROUTER#<COMMAND> ?*—Typing a command followed by a space then a question mark will display any keywords that can be used with that command. For example, typing **CLOCK ?** at the router prompt will return **SET**, the keyword used with the **CLOCK** command.

➤ *ROUTER#<PARTIAL COMMAND>?*—Typing a partial command followed *immediately* by a question mark (no space between partial command and question mark) will display a list of commands that begins with the partial command entry. For example, typing **CL?** at the router prompt will return **CLEAR** and **CLOCK**, the two IOS commands that begin with "CL". Note that if a space is included between a partial or complete command and a question mark, help will interpret this as a keyword request and could result in an "ambiguous command" message. For example, typing **CL ?** (CL, a space, then a question mark) will result in the message "Ambiguous command: "cl " "", because two commands begin with "CL:" (**CLOCK** and **CLEAR**).

For more information, see Chapter 8 of *CCNA Routing and Switching Exam Cram, Second Edition.*

Question 67

The correct answer is a. Pressing the Control key along with the letter *A* will move the cursor to the beginning of the command line. Since this is the only valid use for Ctrl+A, all of the other answers are incorrect. Configuring a Cisco router can sometimes involve long, detailed command lines that can be slow to navigate. For this reason, the Cisco IOS includes a number of "shortcut" ways to navigate the command line. The following list details some of the more common methods used to navigate the command line:

➤ *Ctrl+A*—Pressing the Control key along with the letter *A* will move the cursor to the beginning of the command line.

➤ *Ctrl+E*—Pressing the Control key along with the letter *E* will move the cursor to the end of the command line.

➤ *Ctrl+P*—Pressing the Control key along with the letter *P* will recall previous commands in the command history. (The up arrow will do this as well.)

➤ *Ctrl+N*—Pressing the Control key along with the letter *N* will return to more recent commands in the command history after recalling previous commands. (The down arrow will do this as well.)

➤ *Esc+B*—Pressing the Escape key along with the letter *B* will move the cursor back one word.

➤ *Esc+F*—Pressing the Escape key along with the letter *F* will move the cursor forward one word.

➤ *Left and Right Arrow Keys*—The left and right arrow keys will move the cursor one character left and right, respectively.

➤ *Tab*—Pressing the Tab key will complete an entry typed at the router prompt.

For more information, see Chapter 8 of *CCNA Routing and Switching Exam Cram, Second Edition.*

Question 68

The correct answer is d. **COPY TFTP STARTUP-CONFIG** will copy a configuration file from a TFTP server to the startup configuration in NVRAM. Answer a is incorrect, because the **COPY TFTP STARTUP CONFIG** command uses a hyphen between **STARTUP** and **CONFIG**. Answer b is incorrect, because **COPY STARTUP-CONFIG TFTP** will copy the startup configuration to a TFTP server, not from one. Answer c is incorrect, because the **COPY TFTP-STARTUP-CONFIG** command does not have a hyphen between **TFTP** and **STARTUP**. The startup configuration is a configuration file in a Cisco router that contains the commands used to set router-specific parameters. The startup configuration is stored in NVRAM (Non-Volatile RAM) and is loaded into RAM when the router starts up.

The running configuration is the startup configuration that is loaded into RAM and is the configuration used by the router when it is running. Because any information in RAM is lost when the router is powered off, the running configuration can be saved to the startup configuration, which is located in NVRAM, and will, therefore, be saved when the router is powered off. In addition, both the running and startup configuration files can be copied to and from a TFTP server, for backup purposes. The following list outlines the common commands used when working with configuration files:

➤ *COPY STARTUP-CONFIG RUNNING-CONFIG*—This will copy the startup configuration file in NVRAM to the running configuration in RAM.

➤ *COPY STARTUP-CONFIG TFTP*—This will copy the startup configuration file in NVRAM to a remote TFTP server.

➤ *COPY RUNNING-CONFIG STARTUP-CONFIG*—This will copy the running configuration from RAM to the startup configuration file in NVRAM.

➤ *COPY RUNNING-CONFIG TFTP*—This will copy the running configuration from RAM to a remote TFTP server.

➤ *COPY TFTP RUNNING-CONFIG*—This will copy a configuration file from a TFTP server to the router's running configuration in RAM.

➤ *COPY TFTP STARTUP-CONFIG*—This will copy a configuration file from a TFTP server to the startup configuration in NVRAM.

Note that the basic syntax of the **COPY** command specifies the source first, then the destination.

To modify a router's running configuration, you must be in global configuration mode. The **CONFIG T** command is used in privileged mode to access global configuration mode. Global configuration mode is displayed at the router prompt with "**config**" in parentheses (**RouterA(config)#**). To exit global configuration mode, use the keys Control and Z (Ctrl+Z).

For more information, see Chapter 8 of *CCNA Routing and Switching Exam Cram, Second Edition.*

Question 69

The correct answer is a. The **HOSTNAME** command is used to assign a host name to a Cisco router. All of the other answers are not valid Cisco IOS commands, therefore answers b, c, and d are incorrect. The **HOSTNAME** command must be entered from global configuration mode. When a router is in global configuration mode, the prompt displays "**config**" in parentheses (**Router(config)#**). To display a router's host name, use the **SHOW HOSTS** command in user (**ROUTER>**) or privileged (**ROUTER#**) mode.

For more information, see Chapter 8 of *CCNA Routing and Switching Exam Cram, Second Edition.*

Question 70

The correct answer is e. Access lists can be used to filter the traffic that is handled by a Cisco router. An IP access list will filter IP traffic and can be created as a standard IP access list or an extended IP access list. A standard IP access list can be used to permit or deny access based on IP addressing information only and can only act on the source IP addressing information. An extended IP access list can be used to permit or deny access based on the source and destination IP address information, the protocol, or the port number. Since answers a, b, c, and d can all be specified in an extended IP access list, the best answer to this question is answer e. The syntax for creating an extended IP access list is as follows:

```
ACCESS-LIST [access-list-number] [permit|deny] [protocol] [source]
 [wildcard mask] [destination] [wildmask] [operator][port]
```

➤ *ACCESS-LIST*—This is the command to denote an access list.

➤ *[access-list-number]*—This can be a number between 100 and 199 used to denote an extended IP access list.

➤ *[permit|deny]*—This will allow (permit) or disallow (deny) traffic specified in this access list.

➤ *[protocol]*—This is used to specify the protocol that will be filtered in this access list. Common values are TCP, UDP, and IP. (The use of IP will denote all IP protocols.)

➤ *[source]*—This specifies the source IP addressing information.

➤ *[wildcard mask]*—This can be optionally applied to further define the source. A wildmask can be used to control access to an entire IP network ID rather than a single IP address. A wildcard mask will use the number 255 to mean "any" and the number 0 to mean "must match exactly." The terms **HOST** and **ANY** may also be used here to more quickly specify a single host or an entire network. The **HOST** keyword is the same as the wildcard mask 0.0.0.0. The **ANY** keyword is the same as the wildcard mask 255.255.255.255.

➤ *[destination]*—This specifies the destination IP addressing information.

➤ *[wildmask]*—This can be optionally applied to further define the destination IP addressing.

➤ *[operator]*—This can be optionally applied to define how to interpret the value entered in the **[port]** section of the access list. Common values are **EQ** (equal to the port specified), **LT** (less than the port number specified), and **GT** (greater than the port number specified).

➤ *[port]*—This specifies the port number this access list will act on. Common port numbers include 21 (FTP), 23 (TELNET), 25 (SMTP), 53 (DNS), 69 (TFTP), and 80 (HTTP).

For more information, see Chapter 13 of *CCNA Routing and Switching Exam Cram, Second Edition*.

Practice Test #2

Question 1

Which of the following core components of a Cisco router can permanently
store information after the router is powered off?

○ a. Flash, ROM, and RAM can permanently store information after the
       router is powered off.

○ b. RAM, ROM, and NVRAM can permanently store information after the
       router is powered off.

○ c. ROM, Flash, and NVRAM can permanently store information after the
       router is powered off.

○ d. NVRAM, RAM, and ROM can permanently store information after the
       router is powered off.

Question 2

You are working on your Cisco router in privileged mode and wish to log out.
Which of the following commands can be used to log out of a router from
privileged mode?

○ a. **LOGOUT**

○ b. **LOGOFF**

○ c. **LOG-OFF**

○ d. **LOG-OUT**

Question 3

You are troubleshooting a Cisco router and need to display the configuration information of all interfaces. Which of the following Cisco IOS commands could you issue to view this information?

○ a. **DISPLAY ALL INTERFACES**

○ b. **SHOW ALL**

○ c. **SHOW INTERFACE ALL**

○ d. **SHOW INTERFACE**

Question 4

Your coworker is having trouble copying the running configuration file to the startup configuration on his Cisco router and asks for your assistance. When you go to the router console, you notice the following on the screen:

```
RouterA con0 is now available
Press RETURN to get started.
RouterA>copy running-config startup-config
         ^
% Invalid input detected at '^' marker.
RouterA>
```

Which of the following is the most likely reason that this operation failed?

○ a. The **COPY RUNNING-CONFIG STARTUP-CONFIG** must be entered in all caps, or it will not be executed.

○ b. The **COPY RUNNING-CONFIG STARTUP-CONFIG** must be entered in global configuration mode (RouterA(config)#) or it will not work.

○ c. The correct syntax for this command is **COPY RUNNING-CONFIG STARTUPCONFIG**.

○ d. The **COPY RUNNING-CONFIG STARTUP-CONFIG** must be entered in privileged mode (RouterA#) or it will not work.

Question 5

The TCP/IP configuration of one of the computers in your network is as follows:

```
IP address 208.100.50.90
Subnet Mask 255.255.255.240
```

Based on this information, how many bits, in total, are being used by subnet mask 255.255.255.240?

○ a. 26

○ b. 28

○ c. 24

○ d. 26

Question 6

Seated at the console of your Cisco router, you wish to display the current running configuration. Which of the following is the correct command you will enter to accomplish this?

○ a. **ROUTER> SHOW RUNNING-CONFIG**

○ b. **ROUTER#SHOW RUNNING-CONFIG**

○ c. **ROUTER>SHOW RUN-CONFIG**

○ d. **ROUTER#SHOW RUNNING CONFIG**

Question 7

As the network administrator, you want to require a password to be entered by any user connecting to your Cisco router through the console port. You want the password to be "ADMIN". Which of the following are the correct Cisco IOS commands to perform this operation?

○ a.
```
LINE CON 0
        PASSWORD ADMIN
```

○ b.
```
LINE CON 0
        LOGIN
        PASSWORD ADMIN
```

○ c.
```
LINE VTY 0 4
        LOGIN ADMIN
```

○ d.
```
LINE AUX 0
        LOGIN
        PASSWORD ADMIN
```

Question 8

You manage a network that uses TCP/IP as the only protocol and is divided into multiple subnets. The IP configuration of one of your subnets is as follows:

```
Default gateway = 129.23.98.191
on interface E0
Subnet Mask  = 255.255.255.192
```

Based on this information, what are the valid host IDs for the nodes in this subnet?

○ a. 129.23.98.129 to 129.23.98.190

○ b. 129.23.98.127 to 129.23.98.190

○ c. 129.23.98.128 to 129.23.98.191

○ d. 129.23.98.64 to 129.23.98.190

Question 9

Which of the following protocols reside at the session layer of the OSI model? [Choose the three best answers]

- ❑ a. SMTP
- ❑ b. NFS
- ❑ c. Telnet
- ❑ d. SQL
- ❑ e. PICT
- ❑ f. RPC
- ❑ g. TIFF

Question 10

Consider the following access list:

```
ACCESS LIST 10 PERMIT 200.200.200.75 0.0.0.0
```

Based on this information, which of the following statements best describes the result of this access list?

- ○ a. This access list will allow traffic from host 200.200.200.75.
- ○ b. This access list will allow traffic from host 200.200.200.75 to enter network 0.0.0.0.
- ○ c. This access list will allow traffic from all hosts on network 200.200.200.0.
- ○ d. This access list is invalid.

Question 11

A problem that can occur with distance vector routing protocols is routing loops. One of the methods used by distance vector routing protocols to prevent routing loops is split horizon. Which of the following statements correctly describes the function of split horizon?

○ a.   Split horizon involves setting a maximum hop count that a packet can travel before it is discarded.

○ b.   Split horizon occurs when a router poisons a route by setting a hop count of 16.

○ c.   Split horizon states that a packet cannot be sent back in the direction that it was received.

○ d.   Split horizon is a specified amount of time a router will wait before updating its routing table.

Question 12

You observe the following configuration on one of the computers in your TCP/IP network:

```
IP address : 10.150.33.90
Subnet Mask 255.0.0.0
Default Gateway : 10.22.11.5
```

Based on this information, which of the following statements are true? [Choose the three best answers]

❑ a.   The IP address is a Class C address.

❑ b.   The IP address is a Class A address.

❑ c.   The Network ID of IP address 10.150.33.90 is 10.0.0.0.

❑ d.   The Network ID of IP address 10.150.33.90 is 10.150.0.0.

❑ e.   The Host ID of IP address 10.150.33.90 is 150.33.90.

❑ f.   The Host ID of IP address 10.150.33.90 is 10.150.33.0.

Question 13

You are experiencing problems with the startup configuration on your Cisco router. You have a backup copy of this configuration stored on a TFTP server. Which of the following is the correct Cisco IOS command that you could use to copy a configuration file from your TFTP server to the startup configuration of your router?

○ a. **COPY TFTP STARTUP CONFIG**

○ b. **COPY STARTUP-CONFIG TFTP**

○ c. **COPY TFTP-STARTUP-CONFIG**

○ d. **COPY TFTP STARTUP-CONFIG**

Question 14

Which of the following correctly describes the use of the down arrow at the command line in a Cisco router?

○ a. The down arrow key will move the cursor to the beginning of the command line.

○ b. The down arrow key will recall previous commands in the command history.

○ c. The down arrow key will move the cursor back one word.

○ d. The down arrow key will return to more recent commands in the command history after recalling previous commands.

Question 15

You need to configure a static route on your Cisco router. You want the router to send all packets destined for network 200.200.200.0 255.255.255.0 to router 200.200.200.58. Which of the following is the correct command to create a static route that will perform this operation?

○ a. IP ROUTE 200.200.200.58 255.255.255.0 200.200.200.0

○ b. IP ROUTE 200.200.200.0 255.255.255.0 200.200.200.58

○ c. IP-ROUTE 200.200.200.0 255.255.255.0 200.200.200.58

○ d. IP-ROUTE 200.200.200.58 255.255.255.0 200.200.200.0

Question 16

Which of the following correctly describes the use of Ctrl+E at the command line in a Cisco router?

○ a. Ctrl+E will move the cursor to the beginning of the command line.

○ b. Ctrl+E will move the cursor to the end of the command line.

○ c. Ctrl+E will move the cursor back one word.

○ d. Ctrl+E will move the cursor forward one word.

Question 17

Which of the following best describes the use of the virtual terminal password in a Cisco router?

○ a. The virtual terminal password is used to control access to the Ethernet port on the router.

○ b. The virtual terminal password is used to control remote Telnet access to a router.

○ c. The virtual terminal password is used to control access to any auxiliary ports the router may have.

○ d. The virtual terminal password is used to control access to the privileged mode of the router.

○ e. The virtual terminal password is used to control access to the console port of the router.

○ f. The virtual terminal password is used to control access to the user mode of the router.

Question 18

Which of the following Cisco IOS commands are used to copy the configuration file from a TFTP server into the router's RAM?

○ a. **CONFIGURE TERMINAL (CONFIG T)**

○ b. **CONFIGURE OVERWRITE-NETWORK (CONFIG O)**

○ c. **CONFIGURE NETWORK (CONFIG NET)**

○ d. **CONFIGURE MEMORY (CONFIG MEM)**

Question 19

The IPX/SPX protocol is used in Novell NetWare networks. When IPX packets are passed to the data-link layer, they are encapsulated into frames for transmission over the physical media. Novell NetWare supports a number of different encapsulation methods, or frame types, used to encapsulate IPX packets. The frame type used by NetWare computers must be the same as the one used by a Cisco router, or they will not be able to communicate. When configuring a Cisco router to support IPX, you can specify the frame type to use. Novell NetWare and Cisco both use different names to describe the same frame type. Which of the following statements are true about the frame types used by Novell and Cisco? [Choose the three best answers]

❑ a. The ETHERNET II frame type used by Novell is called ARPA in Cisco.

❑ b. The ETHERNET 802.2 frame type used by Novell is called SNAP in Cisco.

❑ c. The ETHERNET_snap frame type used by Novell is called SNAP in Cisco.

❑ d. The ETHERNET 802.3 frame type used by Novell is called ETHER in Cisco and is the default frame type.

❑ e. The TOKEN RING frame type used by Novell is called ARPA in Cisco.

❑ f. The TOKEN RING_SNAP frame type used by Novell is called SNAP in Cisco.

Question 20

Flow control is a method used by TCP at the transport layer that ensures the amount of data being sent from one node will not overwhelm the destination node. What flow control method temporarily stores packets in memory buffers until they can be processed by the computer?

○ a. Windowing

○ b. Buffering

○ c. Source-quench messages

○ d. Multiplexing

Question 21

Your network contains a NetWare print server. The NetWare print server's network number is 8E.1234.1234.1234 . You do not want your Cisco router to pass any SAP advertisements relating to print services from this server to other networks. To filter this traffic, you decide to create a SAP access list. Which of the following is the correct IPX SAP access list used to perform this operation?

○ a. ACCESS-LIST 1000 DENY 8E.1234.1234.1234 7

○ b. ACCESS-LIST 999 DENY 8E.1234.1234.1234 7

○ c. ACCESS-LIST 1000 DENY 8E.1234.1234.1234 4

○ d. ACCESS LIST 1000 DENY 8E.1234.1234.1234 4

Question 22

At which layer of the OSI model does routing occur?

○ a. Application

○ b. Presentation

○ c. Session

○ d. Transport

○ e. Network

○ f. Data-link

○ g. Physical

Question 23

Seated at the router console, you wish to change the setting for the clock. You are not sure of the command to use, but you recall it does begin with "CL". Which of the following best describes the correct method you could use to find out what Cisco IOS commands begin with the letters "CL"?

○ a. Type "CL?" at the router prompt.

○ b. Type "CL ?" at the router prompt.

○ c. Type "HELP CL/?" at the router prompt.

○ d. Type "SHOW CL*" at the router prompt.

Question 24

ISDN, or Integrated Services Digital Network, is a communications standard that uses digital telephone lines to transmit voice, data, and video. ISDN is a dial-up service that is used on demand. ISDN protocol standards are defined by the International Telecommunication Union (ITU). Which of the following statements are true about ISDN protocol standards?

○ a. Protocols that begin with the letter *E* define ISDN standards on the existing telephone network.

○ b. Protocols that begin with the letter *I* define ISDN standards on the existing telephone network.

○ c. Protocols that begin with the letter *I* specify ISDN switching and signaling standards.

○ d. Protocols that begin with the letter *E* specify ISDN concepts, interfaces, and terminology.

Question 25

Which of the following statements are true about data-link and network layer addresses? [Choose the two best answers]

❑ a. Data-link addresses are considered hierarchical.

❑ b. Network addresses are considered hierarchical.

❑ c. Data-link addresses are considered flat.

❑ d. Network addresses are considered flat.

Question 26

While configuring a Cisco router, you notice that you cannot recall any previously entered commands. You are told that the command history feature has been disabled on this router. What command can you enter on this router to enable the command history feature?

○ a. **ENABLE TERMINAL EDITING**

○ b. **TERMINAL EDITING ENABLE**

○ c. **TERMINAL EDITING ON**

○ d. **TERMINAL EDITING**

Question 27

The bandwidth provided by ISDN can be divided into two categories: Primary Rate ISDN (PRI) and Basic Rate ISDN (BRI). Both Primary and Basic Rate ISDN utilize a D channel when communicating. Which of the following best describes the purpose of the D channel used by ISDN?

○ a. ISDN uses the D channel for control and signaling purposes.

○ b. ISDN uses the D channel to transmit video only.

○ c. ISDN uses the D channel to transmit voice only.

○ d. ISDN uses the D channel to transmit data only.

Question 28

You need to configure an extended IP access list on your Cisco router. You need to control users' access by protocol. Which of the following is not a valid entry for the protocol field in an IP extended access list?

○ a. TCP

○ b. UDP

○ c. IP

○ d. TELNET

○ e. A port number from 0 to 255

Question 29

X.25 is a packet switching wide area network (WAN) technology that utilizes logical, or virtual, circuits to form a connection. X.25 utilizes two basic types of virtual circuits. Which of the following are virtual circuits used in X.25 networks? [Choose the two best answers]

❑ a. Switched virtual circuit (SVC)

❑ b. Permanent virtual circuit (PVC)

❑ c. Local Management Interface (LMI)

❑ d. Data-Link Connection Identifier (DLCI)

Question 30

Which of the following Cisco IOS commands is used to clear the matches for the lines in an access list?

○ a. **CLEAR ACCESS-LIST LINES**

○ b. **CLEAR ACCESS-LIST COUNTERS**

○ c. **CLEAR ACCESS LIST COUNTERS**

○ d. **CLEAR ACCESS-LIST-COUNTERS**

Question 31

Which of the following access lists will allow only host 200.200.200.50 to have Telnet access to network 131.200.0.0?

○ a. ACCESS-LIST 100 PERMIT TCP HOST 200.200.200.50 131.200.0.0 0.0 255.255 EQ 23

○ b. ACCESS-LIST 100 PERMIT TCP 200.200.200.50 255.255.255.0 131.200.0.0 0.0.255.255 EQ 23

○ c. IP ACCESS-LIST 100 PERMIT TCP HOST 200.200.200.50 131.200.0.0 0.0 255.255 EQ 23

○ d. ACCESS-LIST 10 PERMIT TCP HOST 200.200.200.50 131.200.0.0 0.0 255.255 EQ 23

Question 32

You decide to implement a Cisco switch in your network. You want to implement a switching method that offers the least amount of delay as possible. You are not concerned with the error checking of packets before they are forwarded by your switch. Which of the following switching methods will you use in this situation?

○ a. Cut-through switching

○ b. Either store-and-forward switching or cut-through switching

○ c. Store-and-forward switching

○ d. Cut-and-forward switching

Question 33

What three primary sources can be used to load the Cisco IOS (operating system software) during the Cisco router startup sequence? [Choose the three best answers]

❑ a. TFTP server

❑ b. ROM

❑ c. RAM

❑ d. NVRAM

❑ e. Flash

❑ f. Floppy drive

Question 34

Which layer of the OSI model acts as a translator of text and data syntax?

○ a. Application

○ b. Presentation

○ c. Session

○ d. Transport

○ e. Network

○ f. Data-link

○ g. Physical

Question 35

You have configured several IPX access lists on your Cisco router. Which of the following commands can you use to see a list of all the IPX access lists that are currently in use on interface S0 of your Cisco router?

○ a. **SHOW IPX ACCESS-LISTS S0**

○ b. **SHOW IPX INTERFACE S0**

○ c. **SHOW IPX ACCESS-LISTS**

○ d. **SHOW IPX INTERFACE-ACCESS-LISTS S0**

Question 36

What two sublayers is the data-link layer of the OSI model divided into?

○ a. MAC and LLC

○ b. MAC and RPC

○ c. LLC and RPC

○ d. RPC and SQL

Question 37

Which layer of the OSI model is responsible for converting data into bits for transmission by the physical layer?

○ a. Application

○ b. Presentation

○ c. Session

○ d. Transport

○ e. Network

○ f. Data-link

○ g. Physical

Question 38

Encapsulation involves five steps. First, user information is converted to data. Second, data is converted to segments. What is the third step of encapsulation?

○ a. Segments are converted to packets or datagrams.

○ b. Frames are converted to bits.

○ c. Packets and datagrams are converted to frames.

○ d. Segments are converted to data.

Question 39

Which of the following statements is true about protocols and the OSI model?

- ○ a. SMTP, FTP, MIDI, and JPEG are all presentation layer protocols.
- ○ b. FTP, Telnet, EDI, and WAIS are all application layer protocols.
- ○ c. Telnet, WAIS, MIDI, and MPEG are all presentation layer protocols.
- ○ d. WAIS, PICT, MIDI, and Quick Time are all presentation layer protocols.

Question 40

Which layer of the OSI model provides reliable end-to-end communication by utilizing Windowing and Positive Acknowledgments?

- ○ a. Application
- ○ b. Presentation
- ○ c. Session
- ○ d. Transport
- ○ e. Network
- ○ f. Data-link
- ○ g. Physical

Question 41

Consider the following extended IPX access list:

```
ACCESS LIST 922 DENY -1 22 0 37 0
ACCESS LIST 922 PERMIT -1 -1 0 -1 0
```

Based on this access list, which of the following statements is true?

- ○ a. This access list will deny all networks access to IPX network 22 and 37.
- ○ b. This access list will deny IPX network 22 access to IPX network 37.
- ○ c. This access list will deny IPX network 37 access to IPX network 22.
- ○ d. This access list will deny all networks access to IPX network 22, except for IPX network 37.

Question 42

You are troubleshooting a Cisco router that you suspect is not routing information correctly. The router is using TCP/IP as the only protocol. You need to see the contents of the router's routing table. Which of the following Cisco IOS commands can you use to see the routing table?

○ a. **SHOW ROUTING TABLE**

○ b. **SHOW IP ROUTE**

○ c. **SHOW IP-ROUTE**

○ d. **SHOW IP PROTOCOL**

Question 43

Flow control is a method used by TCP at the transport layer that ensures the amount of data being sent from the source node will not overwhelm the destination node. What are the three methods used in flow control? [Choose the three best answers]

❑ a. Windowing

❑ b. Buffering

❑ c. Source-quench messages

❑ d. Multiplexing

Question 44

Which of the following statements correctly describes the procedure used to enter the EXEC privileged mode of a Cisco router?

○ a. From user mode, type the **CONFIGURE** command.

○ b. From user mode, type the **ENABLE** command.

○ c. From system mode, type the **CONFIGURE** command.

○ d. From router mode, type the **ENABLE** command.

Question 45

Which of the following is the correct numerical range used for an extended IP access list?

○ a. 1–99

○ b. 100–199

○ c. 800–899

○ d. 900–999

Question 46

The most common form of WAN connectivity utilizes telecommunication technology. Which of the following best describes the local loop as it is used in WAN communication?

○ a. The local loop is the collection of communication devices (telephones, modems, and so forth) that exists at the customer's location.

○ b. The local loop is the cabling that extends from the DEMARC to the telephone company's Central Office (CO).

○ c. The local loop is the cabling that extends from the ISP to the telephone company's office.

○ d. The local loop is the point at the customer's premises where CPE devices connect and is usually a large punch-down board located in a wiring closet.

Question 47

Which of the following are features of connection-oriented protocols? [Choose the three best answers]

❑ a. Sequencing.

❑ b. Creation of a virtual circuit.

❑ c. A communication session is established between hosts prior to sending data.

❑ d. The guarantee of delivery is the responsibility of higher layer protocols.

Question 48

Which of the following best describes the function and contents of NVRAM (Non-Volatile RAM) in a Cisco router?

- a. NVRAM is a physical chip installed on a router's motherboard that contains the bootstrap program, the Power On Self Test (POST), and the operating system software (Cisco IOS).

- b. NVRAM is used by Cisco routers to store the startup configuration.

- c. NVRAM is an erasable, programmable memory area that contains the Cisco operating system software (IOS).

- d. NVRAM is used as the router's main working area and contains the running configuration.

Question 49

Which of the following best describes the purpose of the BANNER MOTD command?

- a. **BANNER MOTD** is a Cisco IOS command used to configure the Medial Optic Transfer Dilation (MOTD).

- b. **BANNER MOTD** is a Cisco IOS command used to configure a welcome message.

- c. **BANNER MOTD** is a Cisco IOS command used to configure routing protocol headers (banners).

- d. **BANNER MOTD** is a Cisco IOS command used to configure distance vector headers (banners).

Question 50

Routing refers to the process of determining the path a packet will take to its destination. A Cisco router cannot route anything unless an entry is in its routing table that contains information on where to send the packet. A routing table can be of two basic types: static and dynamic. Which of the following statements are true about static and dynamic routing tables? [Choose the two best answers]

- a. Static routing tables are updated using the **IP ROUTE** command.

- b. Dynamic routing tables are updated using the **IP ROUTE** command.

- c. Static routing tables generate more network traffic than dynamic routing tables.

- d. Dynamic routing tables generate more network traffic than static routing tables.

Question 51

Which of the following statements best describes the use of the **CONFIGURE NETWORK (CONFIG NET)** command?

○ a. **CONFIGURE NETWORK** is used to enter configuration commands into the router from the console port or through Telnet.

○ b. **CONFIGURE NETWORK** is used to copy the configuration file from a TFTP server into the router's RAM.

○ c. **CONFIGURE NETWORK** is used to copy a configuration file into NVRAM from a TFTP server.

○ d. **CONFIGURE NETWORK** is used to execute the configuration stored in NVRAM and will copy the startup configuration to the running configuration.

Question 52

Consider the following display on your router console:

```
        — System Configuration Dialog —
At any point you may enter a
question mark '?' for help.
Use ctrl-c to abort configuration
dialog at any prompt.
Default settings are in square brackets'[]'
Continue with configuration
dialog? [yes]: yes
First, would you like to see the
  current interface summary? [yes]: yes
Interface        IP-Address      OK?  Method
Ethernet0        200.200.200.254 YES  NVRAM
Serial0          131.200.50.254  YES  NVRAM
Serial1          unassigned      YES  NVRAM
```

What Cisco IOS command was used to produce this output?

○ a. **SHOW RUNNING-CONFIG**

○ b. **SHOW STARTUP-CONFIG**

○ c. **SETUP**

○ d. **SHOW STARTUP CONFIG**

Question 53

You have configured several IP access lists on your Cisco router. Which of the following commands can you use to see a list of all the IP access lists that are currently in use in your router?

○ a. **SHOW IP ACCESS LIST**

○ b. **DISPLAY IP ACCESS-LIST**

○ c. **SHOW IP ACCESS-LIST-IN-USE**

○ d. **SHOW IP ACCESS-LIST INT E0**

Question 54

Two common protocols used when encapsulating data over serial links are synchronous data-link control (SDLC) and high-level data-link control LC (HD). Which of the following best describes the use of SDLC?

○ a. SDLC is a network layer protocol used by IBM networks when communicating over WAN links using the SNA protocol.

○ b. SDLC is a transport layer protocol used by IBM networks when communicating over WAN links using the SNA protocol.

○ c. SDLC is a data-link layer protocol that is the Cisco default protocol for all serial (WAN) links.

○ d. SDLC is a data-link layer protocol used by IBM networks when communicating over WAN links using the SNA protocol.

Question 55

Which layer of the OSI model is responsible for verifying the appropriate resources are present to initiate a connection?

○ a. Application

○ b. Presentation

○ c. Session

○ d. Transport

○ e. Network

○ f. Data-link

○ g. Physical

Question 56

Which of the following access lists will deny IPX network 15 access to IPX network 25?

○ a. ACCESS-LIST 800 DENY 15 25

○ b. ACCESS-LIST 800 DENY 25 15

○ c. ACCESS-LIST 100 DENY 15 25

○ d. ACCESS LIST 800 DENY 15 25

Question 57

You are configuring IP on serial interface **s1** of your Cisco router. You specify the following at the router console:

```
Configuring interface Serial1:
Is this interface in use? [no]: y
Configure IP on this interface? [no]: y
Configure IP unnumbered on this
interface? [no]: n
IP address for this interface:
 130.200.200.254
Number of bits in subnet field [0]: 7
```

Based on this information, which of the following statements is true?

○ a. This is a Class B address, and the subnet mask is 255.255.254.0.

○ b. This is a Class B address, and the subnet mask is 255.255.248.0.

○ c. This is a Class C address, and the subnet mask is 255.255.255.254.

○ d. This is a Class B address, and the subnet mask is 255.255.0.0.

Question 58

To increase security on your Cisco router, you want to require a password to access the privileged mode of the router. Which of the following passwords can be used to control access to privileged mode on a Cisco router? [Choose the two best answers]

❑ a. The enable password can control access to the privileged mode of a router.

❑ b. The enable secret password can control access to the privileged mode of a router.

❑ c. The virtual terminal password can control access to the privileged mode of a router.

❑ d. The console password can control access to the privileged mode of a router.

❑ e. The auxiliary password can control access to the privileged mode of a router.

❑ f. The privileged password can control access to the privileged mode of a router.

Question 59

Packet switching is a WAN technology that sends data over WAN links through the use of logical circuits, where data is encapsulated into packets and routed by the service provider through various switching points. The two most common packet switching technologies in use are Frame Relay and X.25. which of the following statements are true about Frame Relay and X.25? [Choose the two best answers]

❑ a. The logical connection used in Frame Relay networks is identified by the DLCI.

❑ b. The logical connection used in X.25 networks is identified by the DLCI.

❑ c. The two types of logical circuits used in Frame Relay are SVC and PVC.

❑ d. The two types of logical circuits used in X.25 are SVC and PVC.

Question 60

Both a bridge and a router can be used to segment a network and improve performance. Which of the following statements are true about the difference between a bridge and a router? [Choose the three best answers]

❑ a. A bridge resides at the data-link layer of the OSI model and can filter traffic based on the data-link address of a packet or on a MAC address.

❑ b. A router resides at the data-link layer of the OSI model and can filter traffic based on the data-link address of a packet or on a MAC address.

❑ c. Bridges pass unknown packets out all ports and also pass broadcast traffic.

❑ d. Routers pass unknown packets out all ports and also pass broadcast traffic.

❑ e. Bridges do not pass broadcast traffic and will discard any packets with unknown destinations.

❑ f. Routers do not pass broadcast traffic and will discard any packets with unknown destinations.

Question 61

Which of the following core components of a Cisco router is an erasable, programmable memory area that contains the Cisco operating system software (IOS)?

○ a. Flash

○ b. RAM

○ c. ROM

○ d. Flash RAM

Question 62

Which of the following Exec modes is used to display basic router system information, is used to connect to remote devices, and is limited in capability?

○ a. User mode

○ b. Privileged mode

○ c. Both user mode and privileged mode

○ d. Global configuration mode

Question 63

Which of the following is the correct command and prompt used to enter the **CONFIGURE MEMORY** command?

- ○ a. **ROUTER> CONFIG MEM**
- ○ b. **ROUTER#CONFIG MEM**
- ○ c. **ROUTER(config)#CONFIG MEM**
- ○ d. **ROUTER(CONFIG-IF)#CONFIG MEM**

Question 64

Access lists can be used to filter the traffic that is handled by a Cisco router. Which of the following can be specified in a standard IP access list?

- ○ a. Protocol
- ○ b. Port
- ○ c. Source
- ○ d. Destination
- ○ e. All of the above

Question 65

Which of the following statements best describes the use of the Tab key while entering a Cisco IOS command?

- ○ a. The Tab key will complete an entry typed at the router prompt.
- ○ b. The Tab key will "jump" the router into the next configuration mode.
- ○ c. The Tab key will verify that the syntax of the command you have entered is correct.
- ○ d. The Tab key cannot be used in the Cisco IOS.

Question 66

Which of the following is the correct key(s) used to return to the end of the command line in a Cisco router?

○ a. Ctrl+A

○ b. Ctrl+E

○ c. Ctrl+P

○ d. Ctrl+N

○ e. Esc+B

○ f. Tab

○ g. Right arrow

Question 67

Which of the following is the correct Cisco IOS command used to copy the router's startup configuration file to the running configuration?

○ a. **COPY STARTUP-CONFIG TO RUNNING-CONFIG**

○ b. **COPY STARTUP-CONFIG RUNNING CONFIG**

○ c. **COPY STARTUP-CONFIG RUNNING-CONFIG**

○ d. **COPY RUNNING-CONFIG FROM STARTUP CONFIG**

Question 68

Consider the following display on your router console:

```
Router con0 is now available
Press RETURN to get started.
Router>ENABLE
Password::
```

Based on this display, what type of password could be entered at the Password prompt to continue?

○ a. You must enter the console password to continue.

○ b. You must enter the virtual terminal password to continue.

○ c. You must enter the auxiliary password to continue.

○ d. You must enter the enable password to continue.

Question 69

Which of the following statements is true about IP access lists?

- ○ a. An extended IP access list can be used to permit or deny access based on IP addressing information only and can only act on the source IP addressing information.

- ○ b. An extended access list can be used to permit or deny access based on IP addressing information only and can only act on the source IP addressing information.

- ○ c. An extended IP access list can be used to permit or deny access based on the source and destination IP address, the protocol, or the port number.

- ○ d. A standard IP access list can be used to permit or deny access based on the source and destination IP address, the protocol, or the port number.

Question 70

RIP is a dynamic routing protocol used to automatically update entries in a routing table. Which of the following statements are true about RIP? [Choose the three best answers]

- ❑ a. RIP is a distance vector routing protocol.

- ❑ b. To select the best route, a RIP router selects the path with the fewest number of hops.

- ❑ c. RIP considers hop count, delay, bandwidth, and reliability when determining the best path.

- ❑ d. The maximum hop count used in RIP is 15.

- ❑ e. The maximum hop count used in RIP is 255.

- ❑ f. RIP routers will broadcast the contents of their routing table every 90 seconds.

- ❑ g. RIP is a link state routing protocol.

Answer Key #2

1. c	19. a, c, f	37. f	55. a
2. a	20. b	38. a	56. a
3. d	21. a	39. b	57. a
4. d	22. e	40. d	58. a, b
5. b	23. a	41. b	59. a, d
6. b	24. a	42. b	60. a, c, f
7. b	25. b, c	43. a, b, c	61. a
8. a	26. d	44. b	62. a
9. b, d, f	27. a	45. b	63. b
10. a	28. d	46. b	64. c
11. c	29. a, b	47. a, b, c	65. a
12. b, c, e	30. b	48. b	66. b
13. d	31. a	49. b	67. c
14. d	32. a	50. a, d	68. d
15. b	33. a, b, e	51. b	69. c
16. b	34. b	52. c	70. a, b, d
17. b	35. b	53. a	
18. c	36. a	54. d	

Question 1

The correct answer is c. Read-only memory (ROM), Flash, and Non-Volatile RAM (NVRAM) can permanently store information after the router is powered off. The only core component of a Cisco router that is incapable of storing information when the router is powered off is RAM. The following list outlines the core components of Cisco routers:

➤ *Read-Only Memory (ROM)*—This is a physical chip installed on a router's motherboard that contains the bootstrap program, the Power On Self Test (POST), and the operating system software (Cisco IOS). When a Cisco router is first powered up, the bootstrap program and the POST are executed. ROM cannot be changed through software; the chip must be replaced if any modifications are needed.

➤ *Random Access Memory (RAM)*—As with PCs, RAM is used as the router's main working area and contains the running configuration. All information in RAM dissipates when the router is turned off.

➤ *Non-Volatile RAM (NVRAM)*—This is used by Cisco routers to store the startup or running configuration. After a configuration is created, it exists and runs in RAM. Because RAM cannot permanently store this information, NVRAM is used. Information in NVRAM is preserved after the router is powered off.

➤ *Flash*—Essentially, this is the programmable read-only memory (PROM) used in PCs. The Flash is an erasable, programmable memory area that contains the Cisco operating system software (IOS). When a router's operating system needs to be upgraded, new software can be downloaded from a Trivial File Transfer Protocol (TFTP) server to the router's Flash. Upgrading the IOS in this manner is typically more convenient than replacing the ROM chip in the router's motherboard. Information in Flash is retained when the router is powered off.

For more information, see Chapter 8 of *CCNA Routing and Switching Exam Cram, Second Edition*.

Question 2

The correct answer is a. The **LOGOUT** command is used to log out of a router from privileged mode. The **QUIT**, **EXIT**, and **DISABLE** commands may also be used to log out of privileged mode. All of the other choices are not valid Cisco IOS commands; therefore, answers b, c, and d are incorrect. The Cisco IOS uses a command interpreter known as EXEC, which contains two primary modes: user mode and privileged mode. User mode is used to display basic router system information and to connect to remote devices. The commands in user mode are limited in capability. In user mode, the router prompt is followed by an angle bracket **(ROUTER>)**.

Privileged mode is used to modify and view the configuration of the router, as well as to set IOS parameters. Privileged mode also contains the **CONFIG-URE** command, which is used to access other configuration modes, such as global configuration mode and interface mode. In privileged mode, the router prompt is followed by a pound sign **(ROUTER#)**. To enter privileged mode, enter the **ENABLE** command from user mode.

For more information, see Chapter 8 of *CCNA Routing and Switching Exam Cram, Second Edition.*

Question 3

The correct answer is d. The **SHOW INTERFACE** command is used to display the configuration information of all interfaces. If the **SHOW INTERFACE** command is issued with a specific port (that is, **SHOW INTERFACE E0**), then only the configuration of that interface will be displayed. The **SHOW INTERFACE** command can be issued from user or privileged mode. All of the other choices are not valid Cisco IOS commands; therefore, answers a, b, and c are incorrect.

For more information, see Chapter 8 of *CCNA Routing and Switching Exam Cram, Second Edition.*

Question 4

The correct answer is d. The **COPY RUNNING-CONFIG STARTUP-CONFIG** command will copy the running configuration from RAM to the startup configuration file in NVRAM. It must be entered in privileged mode to execute. In this question, the router is in user mode **(Router>)**. To go into privileged mode, the operator would enter the **ENABLE** command. The prompt would then be followed by a pound sign **(Router#)**. Answer a is incorrect, because the **COPY RUNNING-CONFIG STARTUP-CONFIG** does not have to be entered in all caps. Answer b is incorrect, because the **COPY RUNNING-CONFIG STARTUP-CONFIG** command must be entered in privileged mode, not global configuration mode. Answer c is incorrect, because the **COPY RUNNING-CONFIG STARTUP-CONFIG** was entered with the correct syntax, and the reason it did not work in this question was because it was entered in the wrong mode of the router.

The startup configuration is stored in NVRAM and is loaded into RAM when the router starts up.

The running configuration is the startup configuration that is loaded into RAM and is the configuration used by the router when it is running. Because any information in RAM is lost when the router is powered off, the running configuration can be saved to the startup configuration, which is located in NVRAM and will, therefore, be saved when the router is powered off. In addition, both the running and startup configuration files can be copied to and from a TFTP server for backup purposes. The following list outlines the common commands used when working with configuration files:

➤ *COPY STARTUP-CONFIG RUNNING-CONFIG*—This command will copy the startup configuration file in NVRAM to the running configuration in RAM.

➤ *COPY STARTUP-CONFIG TFTP*—This command will copy the startup configuration file in NVRAM to a remote TFTP server.

➤ *COPY RUNNING-CONFIG STARTUP-CONFIG*—This command will copy the running configuration from RAM to the startup configuration file in NVRAM.

➤ *COPY RUNNING-CONFIG TFTP*—This command will copy the running configuration from RAM to a remote TFTP server.

➤ *COPY TFTP RUNNING-CONFIG* — This command will copy a configuration file from a TFTP server to the router's running configuration in RAM.

➤ *COPY TFTP STARTUP-CONFIG*—This command will copy a configuration file from a TFTP server to the startup configuration in NVRAM.

Note that the basic syntax of the **COPY** command specifies the source first, then the destination.

For more information, see Chapter 8 of *CCNA Routing and Switching Exam Cram, Second Edition*.

Question 5

The correct answer is b. The subnet mask, 255.255.255.240, uses a total of 28 bits. An IP address is a 32-bit network layer address that comprises a network ID and a host ID. A subnet mask is used to distinguish the network portion of an IP address from the host portion. The 32 bits in an IP address are expressed in dotted decimal notation in four octets (four 8-bit numbers separated by periods, that is, 208.100.50.90). The subnet mask, 255.255.255.240, is said to use 28 bits, because it takes 28 of 32 bit positions to create. And 255.255.255.240 can be expressed in binary as 11111111.11111111.11111111.11110000, which is three octets using 8 bits each (24) and 4 bits used in the fourth octet to make the number 240. Because the subnet mask is using 28 bits, and an IP address is a total of 32 bits, 4 bits remain to create host IDs on this network.

For more information, see Chapter 13 of *CCNA Routing and Switching Exam Cram, Second Edition.*

Question 6

The correct answer is b. The **SHOW RUNNING-CONFIG** command can be used in privileged mode to display the current running configuration in RAM. Answers a and c are incorrect, because the router is in user mode, which means the router prompt is followed by an angle bracket (**ROUTER>**). **SHOW RUNNING-CONFIG** cannot be used in either user or global configuration modes. In privileged mode, the router prompt is followed by a pound sign (**ROUTER#**). Answer d is incorrect, because the **SHOW RUNNING-CONFIG** contains a hyphen between **RUNNING** and **CONFIG**.

For more information, see Chapter 8 of *CCNA Routing and Switching Exam Cram, Second Edition.*

Question 7

The correct answer is b. The console password is used to control access to the console port of the router. Setting the console password will require all users connecting to the router console to provide this password for access. To configure the console password, use the following commands:

```
LINE CON 0
LOGIN
PASSWORD [password]
```

Answer a is incorrect, because the **LOGIN** command is missing. Answer c is incorrect, because **LINE VTY** is used to set the virtual terminal password, not the console password. Answer d is incorrect, because **LINE AUX** is used to set the auxiliary password, not the console password.

Cisco routers utilize passwords to provide security access to a router. Passwords can be set for controlling access to privileged mode, for remote sessions via Telnet, or for the auxiliary port. The following list details the common passwords used in Cisco routers:

➤ *Enable Password*—The enable password is used to control access to the privileged mode of the router. If an enable password is specified, it must be entered after the **ENABLE** command to successfully access privileged mode. To configure the enable password, use the **ENABLE PASSWORD** [*password*] command.

➤ *Enable Secret Password*—The enable secret password is used to control access to privileged mode, similar to the enable password. The difference between enable secret and enable password is the enable secret password will be encrypted for additional security, and it will take precedence over the enable password if both are enabled. To configure the enable secret password, use the **ENABLE SECRET** [*password*] command.

➤ *Virtual Terminal Password*—The virtual terminal password is used to control remote Telnet access to a router. Setting the virtual terminal password will require all users that Telnet into the router to provide this password for access. To configure the virtual terminal password, use the following commands:

```
LINE VTY 0 4
LOGIN
PASSWORD [password]
```

➤ *Auxiliary Password*—The auxiliary password is used to control access to any auxiliary ports the router may have. Setting the auxiliary password will require all users connecting to the auxiliary port of the router (usually remote dial-in) to provide this password for access. To configure the auxiliary password, use the following commands:

```
LINE AUX 0
LOGIN
PASSWORD [password]
```

➤ *Console Password*—The console password is used to control access to the console port of the router. Setting the console password will require all users connecting to the router console to provide this password for access. To configure the console password, use the following commands:

```
LINE CON 0
LOGIN
PASSWORD [password]
```

For more information, see Chapter 8 of *CCNA Routing and Switching Exam Cram, Second Edition.*

Question 8

The correct answer is a. The IP address range, 129.23.98.129 to 129.23.98.190, are valid host IDs for the nodes in this subnet. When subnetting IP addresses, host IDs are grouped together according to the incremental value of the subnet mask in use. The first address in this range identifies the subnet (subnet ID) and is not used for a host ID. The last address in this range is used by the router as the broadcast address, also known as the *default gateway*. In this question, the subnet mask used is 192, which will divide subnets in groups of 64 (129.23.98.64 through 129.23.98.127 and 129.23.98.128 through 129.23.98.191). The default gateway is specified as 129.23.98.191, which places this address in subnet 129.23.98.128 through 129.23.98.191. The first address is used as the subnet ID and the last address is used for the default gateway, making the valid range of host IDs 129.23.98.129 to 129.23.98.190. The following list details the incremental values of IP subnet masks:

➤ *Subnet mask 192*—Increments in 64

➤ *Subnet mask 224*—Increments in 32

➤ *Subnet mask 240*—Increments in 16

➤ *Subnet mask 248*—Increments in 8

➤ *Subnet mask 252*—Increments in 4

➤ *Subnet mask 254*—Increments in 2

➤ *Subnet mask 255*—Increments in 1

For more information, see Chapter 8 of *CCNA Routing and Switching Exam Cram, Second Edition.*

Question 9

The correct answers are b, d, and f. The session layer of the OSI model is used to coordinate communication between nodes and is also responsible for creating, maintaining, and ending a communication session. Session layer protocols include Network File System (NFS, used by SUN Microsystems and Unix with TCP/IP), structured query language (SQL, used to define database information requests), remote procedure calls (RPC, used in Microsoft network communication), X Window (used by Unix terminals), AppleTalk Session Protocol (ASP, used by Apple computers), and Digital Network Architecture

Session Control Protocol (DNA SCP, used by IBM). Answers a and c are incorrect, because Simple Mail Transfer Protocol (SMTP) and Telnet are application layer protocols. Answers e and g are incorrect, because PICTure (PICT, Apple computer picture format) and Tagged Image File Format (TIFF) are presentation layer protocols.

For more information, see Chapter 4 of *CCNA Routing and Switching Exam Cram, Second Edition.*

Question 10

The correct answer is a. This access list will allow traffic from host 200.200.200.75. Answer b is incorrect, because a wildmask is not used to specify a network ID. Answer c is incorrect, because the wildmask, 0.0.0.0, will only allow a single host access, not an entire network ID. Answer d is incorrect, because this is a valid access list.

Access lists can be used to filter the traffic that is handled by a Cisco router. An IP access list will filter IP traffic and can be created as a standard IP access list or an extended IP access list. A standard IP access list can be used to permit or deny access based on IP addressing information only and can only act on the source IP addressing information. The syntax for creating a standard IP access list is:

```
ACCESS-LIST [access-list-number]
[permit|deny] source [wildcard mask]
```

➤ *ACCESS-LIST*—This is a command followed by the **access list number**. When using a standard IP access list, this can be a number between 1 and 99.

➤ *[permit|deny]*—This statement will either allow or disallow traffic in this access list.

➤ *source*—This statement is used to specify the source IP addressing information. A standard IP access list can only act on a source address.

➤ *[wildcard mask]*—This is an optional parameter in a standard IP access list. If no wildcard mask is specified, the access list will only act on a single IP address that is specified in the **source**. A wildmask can be used to control access to an entire IP network ID rather than a single IP address. A **wildcard mask** will use the number 255 to mean "any" and the number 0 to mean "must match exactly."

For more information, see Chapter 13 of *CCNA Routing and Switching Exam Cram, Second Edition.*

Question 11

The correct answer is c. Split horizon states that a packet cannot be sent back in the direction that it was received. Answer a is incorrect, because maximum hop count involves setting a maximum hop count that a packet can travel before it is discarded. Answer b is incorrect, because route poisoning occurs when a router poisons a route by setting a hop count of 16. Answer d is incorrect, because a hold-down is a specified amount of time a router will wait before updating its routing table.

A routing loop occurs when a link on a router fails, and some routers do not receive the update information. The routers that do not have the updated information in their routing table will advertise to other routers that they have a route to the failed link and will begin receiving the packets destined for that network. Because this route is failed and there is no real way to route the packet to its destination, packets will end up getting passed back and forth between routers and never reach their destinations. To combat the problem of routing loops, distance vector routing protocols utilize four basic techniques: maximum hop count, split horizon, route poisoning, and hold-downs. The following list describes the function of each:

➤ *Maximum Hop Count*—This involves setting a maximum hop count that a packet can travel before it is discarded. RIP uses a maximum hop count of 15, so that in the event of a routing loop, a packet that travels more than 15 hops will be discarded.

➤ *Split Horizon*—This states that a packet cannot be sent back in the direction from which it was received. This prevents the routing loop of two routers passing a packet back and forth between each other.

➤ *Route Poisoning*—When a router that is using a distance vector routing protocol has a link that fails, it will "poison" its route by setting the hop count of the failed link to a number over the maximum hop count, so that other routers will deem the route unreachable. In the case of RIP, a router can poison a route by setting a hop count of 16. (Fifteen is the maximum hop count used by RIP.)

➤ *Hold-Downs*—This is a specified amount of time a router will wait before updating its routing table. Hold-downs prevent routing loops by preventing routers from updating new information too quickly, allowing time for a failed link to be reestablished.

For more information, see Chapter 11 of *CCNA Routing and Switching Exam Cram, Second Edition*.

Question 12

The correct answers are b, c, and e. IP address 10.150.33.90 is a Class A address with a network ID of 10.0.0.0 and a host ID of 150.33.90. An IP address is a 32-bit network layer address that comprises a network ID and a host ID. TCP/IP addresses are divided into three main classes: Class A, Class B, and Class C. The "first octet" rule specifies that an IP address can be identified by class according to the number in the first octet. The address range for each address class is as follows:

➤ *Class A*—These are IP addresses with numbers from 1 to 126 in the first octet. (Note: 127 is not used as a network ID in IP; it is reserved for localhost loopback testing.)

➤ *Class B*—These are IP addresses with numbers from 128 to 191 in the first octet.

➤ *Class C*—These are IP addresses with numbers from 192 to 223 in the first octet.

When looking at any IP address, it is the number in the *first* octet that determines the address class of that address. For example:

➤ The IP address 3.100.32.7 is a Class A address, because it *begins* with 3, a number between 1 and 126.

➤ The IP address 145.100.32.7 is a Class B address, because it *begins* with 145, a number between 128 and 191.

➤ The IP address 210.100.32.7 is a Class C address, because it *begins* with 210, a number between 192 and 223.

An IP address is made up of a network ID and a host ID. The network ID identifies the network a node is on, and the host ID identifies the individual node. An IP address is specified using four individual numbers separated by three periods. Each one of these "sections" is known as an octet (because it contains 8 bits). Each address class uses different combinations of octets to specify a network ID and host ID. The configuration is as follows:

➤ *Class A*—This uses the first octet for the network ID and the last three octets for the host ID.

> ➤ IP address 2.3.45.9 is a Class A address (it begins with 2, a number between 1 and 126). The network ID of this IP address is 2, and the host ID is 3.45.9, because a Class A address uses the first octet for the network ID and the last three octets for the host ID.

➤ *Class B*—This uses the first two octets for the network ID and the last two octets for the host ID.

   ➤ IP address 167.211.32.5 is a Class B address (it begins with 167, a number between 128 and 191). The network ID of this IP address is 167.211, and the host ID is 32.5, because a Class B address uses the first two octets for the network ID and the last two octets for the host ID.

➤ *Class C*—This uses the first three octets for the network ID and the last octet for the host ID.

   ➤ IP address 200.3.11.75 is a Class C address (it begins with 200, a number between 192 and 223). The network ID of this IP address is 200.3.11 and the host ID is 75, because a Class C address uses the first three octets for the network ID and the last octet for the host ID.

For more information, see Chapter 8 of *CCNA Routing and Switching Exam Cram, Second Edition.*

Question 13

The correct answer is d. The **COPY TFTP STARTUP-CONFIG** command will copy a configuration file from a TFTP server to the startup configuration in NVRAM. Answer a is incorrect, because the correct syntax of the **COPY TFTP STARTUP-CONFIG** command has a hyphen between **STARTUP** and **CONFIG**. Answer b is incorrect, because **COPY STARTUP-CONFIG TFTP** will copy the startup configuration file in NVRAM to a remote TFTP server. Answer c is incorrect, because the correct syntax of the **COPY TFTP STARTUP-CONFIG** command has a hyphen between **STARTUP** and **CONFIG** but no hyphen between **TFTP** and **STARTUP**.

The startup configuration is a configuration file in a Cisco router that contains the commands used to set router-specific parameters. The startup configuration is stored in NVRAM (Non-Volatile RAM) and is loaded into RAM when the router starts up.

The running configuration is the startup configuration that is loaded into RAM and is the configuration used by the router when it is running. Because any information in RAM is lost when the router is powered off, the running configuration can be saved to the startup configuration, which is located in NVRAM and will, therefore, be saved when the router is powered off. In addition, both the running and startup configuration files can be copied to and from a TFTP server for backup purposes. The following list outlines the common commands used when working with configuration files:

➤ *COPY STARTUP-CONFIG RUNNING-CONFIG*—This command will copy the startup configuration file in NVRAM to the running configuration in RAM.

➤ *COPY STARTUP-CONFIG TFTP*—This command will copy the startup configuration file in NVRAM to a remote TFTP server.

➤ *COPY RUNNING-CONFIG STARTUP-CONFIG*—This command will copy the running configuration from RAM to the startup configuration file in NVRAM.

➤ *COPY RUNNING-CONFIG TFTP*—This command will copy the running configuration from RAM to a remote TFTP server.

➤ *COPY TFTP RUNNING-CONFIG*—This command will copy a configuration file from a TFTP server to the router's running configuration in RAM.

➤ *COPY TFTP STARTUP-CONFIG*—This command will copy a configuration file from a TFTP server to the startup configuration in NVRAM.

Note that the basic syntax of the **COPY** command specifies the source first, then the destination.

To modify a router's running configuration, you must be in global configuration mode. The **CONFIG T** command is used in privileged mode to access global configuration mode. Global configuration mode is displayed at the router prompt with "**config**" in parentheses **(RouterA(config)#)**. To exit global configuration mode, use the keys Control and Z (Ctrl+Z).

For more information, see Chapter 8 of *CCNA Routing and Switching Exam Cram, Second Edition*.

Question 14

The correct answer is d. Pressing the down arrow key will return to more recent commands in the command history after recalling previous commands (Ctrl+N will do this as well). Since this is the only function of the down arrow key, all of the other answers are incorrect. Configuring a Cisco router can sometimes involve long, detailed command lines that can be slow to navigate. For this reason, the Cisco IOS includes a number of shortcuts through them. The following list details some of the more common methods used to navigate the command line:

➤ *Ctrl+A*—Pressing the Control key along with the letter *A* will move the cursor to the beginning of the command line.

➤ *Ctrl+E*—Pressing the Control key along with the letter *E* will move the cursor to the end of the command line.

➤ *Ctrl+P*—Pressing the Control key along with the letter *P* will recall previous commands in the command history. (The up arrow will do this as well.)

➤ *Ctrl+N*—Pressing the Control key along with the letter *N* will return to more recent commands in the command history after recalling previous commands. (The down arrow will do this as well.)

➤ *Esc+B*—Pressing the Escape key along with the letter *B* will move the cursor back one word.

➤ *Esc+F*—Pressing the Escape key along with the letter *F* will move the cursor forward one word.

➤ *Left and Right Arrow Keys*—The left and right arrow keys will move the cursor one character left and right, respectively.

➤ *Tab*—Pressing the Tab key will complete an entry typed at the router prompt.

For more information, see Chapter 8 of *CCNA Routing and Switching Exam Cram, Second Edition.*

Question 15

The correct answer is b. The **IP ROUTE 200.200.200.0 255.255.255.0 200.200.200.58** command will create a static route in the router's routing table that will send all packets destined for network **200.200.200.0 255.255.255.0** to router **200.200.200.58**. Answer a is incorrect, because the **IP ROUTE** command specifies the source network first, then the destination router. Answers c and d are incorrect, because the **IP ROUTE** command does not have a hyphen between **IP** and **ROUTE**. A routing table can be of two basic types: static and dynamic. A static routing table is one that has its entries entered manually by an operator. On a Cisco router, a static route can be entered using the **IP ROUTE** command. The syntax for the **IP ROUTE** command is as follows:

```
IP ROUTE [network] [subnet mask] [IP address] [ distance]
```

➤ *IP ROUTE*—This command signifies this will be a static entry in the routing table.

➤ *[network]*—This is the destination network ID.

➤ *[subnet mask]*—This is the subnet mask of the destination network.

➤ *[IP address]*—This is the IP address of the router that will receive all packets that have the address specified in the **NETWORK** statement.

➤ *[distance]*—This is an optional parameter that is the administrative distance of this route. It tells the router the relative importance of this route. If a router has multiple entries in its routing table for the same network ID, it will use the route with the lowest administrative distance. The default administrative distance for a static route is 1. (The number 1 is the second highest priority for administrative distance. Zero is the highest and signifies a directly connected interface.)

For more information, see Chapter 11 of *CCNA Routing and Switching Exam Cram, Second Edition*.

Question 16

The correct answer is b. Pressing the Control key along with the letter *E* will move the cursor to the end of the command line. Because this is the only use of the Ctrl+E key, all of the other answers are incorrect. Configuring a Cisco router can sometimes involve long, detailed command lines that can be slow to navigate. For this reason, the Cisco IOS includes a number of shortcuts through them. The following list details some of the more common methods used to navigate the command line:

➤ *Ctrl+A*—Pressing the Control key along with the letter *A* will move the cursor to the beginning of the command line.

➤ *Ctrl+E*—Pressing the Control key along with the letter *E* will move the cursor to the end of the command line.

➤ *Ctrl+P*—Pressing the Control key along with the letter *P* will recall previous commands in the command history. (The up arrow will do this as well.)

➤ *Ctrl+N*—Pressing the Control key along with the letter *N* will return to more recent commands in the command history after recalling previous commands. (The down arrow will do this as well.)

➤ *Esc+B*—Pressing the Escape key along with the letter *B* will move the cursor back one word.

➤ *Esc+F*—Pressing the Escape key along with the letter *F* will move the cursor forward one word.

➤ *Left and Right Arrow Keys*—These arrow keys will move the cursor one character left and right, respectively.

➤ *Tab*—Pressing the Tab key will complete an entry typed at the router prompt.

For more information, see Chapter 8 of *CCNA Routing and Switching Exam Cram, Second Edition.*

Question 17

The correct answer is b. The virtual terminal password is used to control remote Telnet access to a router. Answer a is incorrect, because there is no password to control access to the Ethernet port on a Cisco router. Answer c is incorrect, because the auxiliary password is used to control access to the auxiliary port on a Cisco router. Answer d is incorrect, because the enable or enable secret passwords are used to control access to the privileged mode of the router. Cisco routers utilize passwords to provide security access to a router. Passwords can be set for controlling access to privileged mode, to remote sessions via Telnet, or to the auxiliary port. The following list details the common passwords used in Cisco routers:

➤ *Enable Password*—The enable password is used to control access to the privileged mode of the router. If an enable password is specified, it must be entered after the **ENABLE** command to successfully access privileged mode. To configure the enable password, use the **ENABLE PASS-WORD** [*password*] command.

➤ *Enable Secret Password*—The enable secret password is used to control access to privileged mode, similar to the enable password. The difference between enable secret and enable password is that the enable secret password will be encrypted for additional security, and it will take precedence over the enable password if both are enabled. To configure the enable secret password, use the **ENABLE SECRET** [**password**] command.

➤ *Virtual Terminal Password*—The virtual terminal password is used to control remote Telnet access to a router. Setting the virtual terminal password will require all users that Telnet into the router to provide this password for access. To configure the virtual terminal password, use the following commands:

```
LINE VTY 0 4
LOGIN
PASSWORD [password]
```

➤ *Auxiliary Password*—The auxiliary password is used to control access to any auxiliary ports the router may have. Setting the auxiliary password will require all users connecting to the auxiliary port of the router (usually remote dial-in) to provide this password for access. To configure the auxiliary password, use the following commands:

```
LINE AUX 0
LOGIN
PASSWORD [password]
```

➤ *Console Password*—The console password is used to control access to the console port of the router. Setting the console password will require all users connecting to the router console to provide this password for access. To configure the console password, use the following commands:

```
LINE CON 0
LOGIN
PASSWORD [password]
```

For more information, see Chapter 8 of *CCNA Routing and Switching Exam Cram, Second Edition.*

Question 18

The correct answer is c. **CONFIGURE NETWORK (CONFIG NET)** is used to copy the configuration file from a TFTP server into the router's RAM. Answer a is incorrect, because the **CONFIGURE TERMINAL** command is used to enter configuration commands into the router from the console port or through Telnet. Answer b is incorrect, because the **CONFIGURE NETWORK (CONFIG NET)** command is used to copy the configuration file from a TFTP server into the router's RAM. Answer d is incorrect, because the **CONFIGURE MEMORY (CONFIG MEM)** command is used to execute the configuration stored in NVRAM. The **CONFIGURE** command, executed in privileged mode, is used to enter the configuration mode of a Cisco router. The four parameters used with the **CONFIGURE** command are **TERMINAL, MEMORY, OVER-WRITE-NETWORK,** and **NETWORK.** The following list outlines the function of these parameters:

➤ *CONFIGURE TERMINAL (CONFIG T)*—This is used to enter configuration commands into the router from the console port or through Telnet.

➤ *CONFIGURE NETWORK (CONFIG NET)*—This is used to copy the configuration file from a TFTP server into the router's RAM.

➤ *CONFIGURE OVERWRITE-NETWORK (CONFIG O)*—This is used to copy a configuration file into NVRAM from a TFTP server. Use of the **CONFIGURE OVERWRITE-NETWORK** will not alter the running configuration.

➤ *CONFIGURE MEMORY (CONFIG MEM)*—This is used to execute the configuration stored in NVRAM. It will copy the startup configuration to the running configuration.

For more information, see Chapter 8 of *CCNA Routing and Switching Exam Cram, Second Edition.*

Question 19

The correct answers are a, c, and f. The IPX/SPX protocol is used in Novell NetWare networks. When IPX packets are passed to the data-link layer, they are encapsulated into frames for transmission over the physical media. Novell NetWare supports a number of different encapsulation methods, or frame types, used to encapsulate IPX packets. The frame type used by NetWare computers must be the same as the one used by a Cisco router, or they will not be able to communicate. When configuring a Cisco router to support IPX, you can specify the frame type to use. Novell NetWare and Cisco both use different names to describe the same frame type. Answer b is incorrect, because the ETHERNET 802.2 frame type is called SAP in Cisco. Answer d is incorrect, because the ETHERNET 802.3 frame type is called NOVELL-ETHER in Cisco. Answer e is incorrect, because the TOKEN RING frame type is called TOKEN in Cisco. The following list details the frame types used by NetWare networks and the comparable Cisco name:

➤ *ETHERNET II*—This is used by Novell and is called ARPA in Cisco.

➤ *ETHERNET 802.2*—This is used by Novell and is called SAP in Cisco.

➤ *ETHERNET_snap*—This is used by Novell and is called SNAP in Cisco.

➤ *ETHERNET 802.3*—This is used by Novell and is called NOVELL-ETHER in Cisco and is the default frame type.

➤ TOKEN RING—This is used by Novell and is called TOKEN in Cisco.

➤ *TOKEN RING_SNAP*—This is used by Novell and is called SNAP in Cisco.

For more information, see Chapter 12 of *CCNA Routing and Switching Exam Cram, Second Edition.*

Question 20

The correct answer is b. Buffering is a method of temporarily storing packets in memory buffers until they can be processed by the computer. Answer a is incorrect, because **Windowing** is the process of predetermining the amount of data that will be sent before an acknowledgment is expected. Answer c is incorrect, because **Source-Quench Messaging** is a message sent by a receiving computer when its memory buffers are nearing capacity. Answer d is incorrect, because **Multiplexing** is the ability of multiple connection bandwidths to utilize a single transport and is not used as part of flow control. Three basic techniques are used in flow control:

➤ *Windowing*—This is the process of predetermining the amount of data that will be sent before an acknowledgment is expected.

➤ *Buffering*—This is a method of temporarily storing packets in memory buffers until they can be processed by the computer.

➤ *Source-Quench Messaging*—A message sent by a receiving computer when its memory buffers are nearing capacity. A source-quench message is a device's way of telling the sending computer to slow down the rate of transmission.

➤ *Multiplexing*—This is the ability of multiple connection bandwidths to utilize a single transport and is not used as part of flow control.

For more information, see Chapter 9 of *CCNA Routing and Switching Exam Cram, Second Edition*.

Question 21

The correct answer is a. The access list **ACCESS-LIST 1000 DENY 8E.1234.1234.1234 7** will deny the advertisement of print services from NetWare server **8E.1234.1234.1234 4**. Answer b is incorrect, because an IPX SAP access list uses numbers between 1000 and 1099. Answer c is incorrect, because the service number is 4, which denotes a NetWare file server. Answer d is incorrect, because no hyphen is between **ACCESS** and **LIST**. An IPX SAP filter can be applied to a router's interface to filter the Service Advertising Protocol (SAP) advertisement traffic passed by the router. Novell NetWare servers will broadcast a list of their available resources every 60 seconds. This is known as a SAP advertisement. Cisco routers will record this information into their own SAP table for advertisement to other segments. By filtering the amount of SAP traffic with an IPX SAP access list, network traffic can be controlled and/or reduced. The syntax for a SAP access list is as follows:

```
ACCESS-LIST [access-list-number] [permit|deny] [source network]
[service type]
```

➤ *ACCESS-LIST*—This command is followed by the **access list number.** When using an IPX SAP access list, this can be a number between 1000 and 1099.

➤ *[permit|deny]*—This statement will either allow or disallow traffic in this access list.

➤ *[source network]*—This statement is used to specify the source IPX network. A "-1" can be used to signify "all networks."

➤ *[service type]*—This statement is used to specify the type of service to be filtered. Common IPX service types include a NetWare File server, specified as number 4, and a NetWare printer server, specified as number 7.

For more information, see Chapter 13 of *CCNA Routing and Switching Exam Cram, Second Edition.*

Question 22

The correct answer is e. The OSI model defines standards used in networking and comprises a seven-layer model. The network layer of the OSI model is responsible for routing and makes use of logical addresses to determine the path a packet will take to get to its destination. Since no other layer of the OSI model is responsible for routing, all of the other answers are incorrect. The following list outlines the seven layers of the OSI model and their functions:

➤ *Application*—The "window" to networking used by programs, the application layer is responsible for:

➤ Verifying that the appropriate resources are present to initiate a connection with the destination node.

➤ Verifying the identity of the destination node.

Application layer protocols include Telnet, File Transfer Protocol (FTP), Simple Mail Transfer Protocol (SMTP), World Wide Web (WWW), Electronic Data Interchange (EDI), and Wide Area Information Server (WAIS).

➤ *Presentation*—Essentially a translator, the presentation layer is responsible for:

➤ Translating text and data syntax, such as Extended Binary-Coded Decimal Interchange Code (EBCDIC, used in IBM systems) to

American Standard Code for Information Interchange (ASCII, used in PC and most computer systems).

➤ Using Abstract Syntax Notation to perform data translation.

Presentation layer protocols include PICTure (PICT, Apple computer picture format), Tagged Image File Format (TIFF), Joint Photographic Experts Group (JPEG), Musical Instrument Digital Interface (MIDI), Motion Picture Expert Group (MPEG), and Quick Time (audio/video application).

➤ *Session*—This is used to coordinate communication between nodes; the session layer is responsible for:

➤ Creating, maintaining, and ending a communication session.

➤ Coordinating service requests and responses that occur between nodes across the network.

Session layer protocols include Network File System (NFS, used by SUN Microsystems and Unix with TCP/IP), structured query language (SQL, used to define database information requests), Remote Procedure Calls (RPC, used in Microsoft network communication), X Window (used by Unix terminals), AppleTalk Session Protocol (ASP, used by Apple computers), and Digital Network Architecture Session Control Protocol (DNA SCP, used by IBM).

➤ *Transport*—Reliable end-to-end communication is the primary function of the transport layer. Its many responsibilities include:

➤ Ensuring flow control so the amount of data being sent from one node will not overwhelm the destination node.

➤ Ensuring the ability of multiple applications to utilize a single transport (multiplexing).

➤ Ensuring the reliable transfer of data. The transport layer implements connection-oriented services between nodes, which utilizes a three-way handshake (synchronization, acknowledgment, and data transfer) to efficiently transfer data.

➤ Ensuring positive acknowledgment, which is the process of a node waiting for an acknowledgment from the destination node prior to sending data.

➤ Windowing, which is a form of flow control that specifies how much data will be transferred between acknowledgments.

Transport layer protocols include Transmission Control Protocol (TCP, part of the TCP/IP protocol suite), and Sequenced Packet Exchange (SPX, part of the IPX/SPX protocol suite).

➤ *Network*—This selects the appropriate path a packet should take to get to its intended destination. It is responsible for routing, which is the process of using a network layer address to determine the best path a packet will travel to its destination. Network layer protocols include Internet Protocol (IP, part of the TCP/IP protocol suite), and Internetwork Packet Exchange (IPX, part of the IPX/SPX protocol suite).

➤ *Data Link*—Divided into two sublayers (Media Access Control [MAC], and Logical Link Control [LLC]), the data-link layer is responsible for:

➤ Preparing data from upper layers to be transmitted over the physical medium by encapsulating upper-layer data into frames. This frame includes the source and destination of MAC addresses in the frame header.

➤ Converting data into bits, so it can be transmitted by the physical layer.

➤ Adding a cyclical redundancy check (CRC) to the end of a frame, used for error checking at the data-link layer.

Data-link protocols include High-Level Data Link Control (HDLC, the Cisco default encapsulation for serial connections), Synchronous Data Link Control, (SDLC, used in IBM networks), Link Access Procedure Balanced (LAPB, used with X.25), X.25 (packet switching network), Serial Line Internet Protocol (SLIP, an older TCP/IP dial-up protocol), Point-to-Point Protocol (PPP, a newer dial-up protocol), Integrated Services Digital Network (ISDN, a dial-up digital service), and frame relay (a packet switching network).

➤ *Physical*—The lowest layer of the OSI model, the physical layer defines the electrical functionality required to send and receive bits over a given physical medium. Specifications that define the voltage levels and physical components of a network are defined at the physical layer. Protocols are not specified at the physical layer, because they are implemented as software. Examples of the standards for sending data over the physical medium are Ethernet (the most widely used standard in networking), Token Ring (IBM's proprietary networking topology), and Fiber Distributed Data Interface (FDDI, a standard for fiber optic networks, commonly used as a backbone).

For more information, see Chapter 4 of *CCNA Routing and Switching Exam Cram, Second Edition.*

Question 23

The correct answer is a. Typing a partial command followed *immediately* by a question mark (no space between partial command and question mark) will display a list of commands that begin with the partial command entry. Typing CL? at the router prompt will return **CLEAR** and **CLOCK**, the two IOS commands that begin with "CL". Answer b is incorrect, because there is a space between "CL" and the question mark. If a space is included between a partial or complete command and a question mark, help will interpret this as a keyword request and could result in an "ambiguous command" message. Typing "CL ?" (CL, a space, then a question mark) will result in the message "Ambiguous command: "cl " ", because two commands begin with "CL": (**CLOCK** and **CLEAR**). Answers c and d are not valid Cisco IOS commands and, therefore, are incorrect.

For more information, see Chapter 8 of *CCNA Routing and Switching Exam Cram, Second Edition.*

Question 24

The correct answer is a. ISDN is a communications standard that uses digital telephone lines to transmit voice, data, and video. It is a dial-up service that is used on demand, and its protocol standards are defined by the International Telecommunication Union (ITU). Answers b and c are incorrect, because protocols that begin with the letter *I* (such as I.430 and I.431) specify ISDN concepts, interfaces, and terminology. Answer d is incorrect, because protocols that begin with the letter *E* (such as E.164) define ISDN standards on the existing telephone network. Protocols that begin with the letter *Q* (such as Q.931 and Q.921) specify ISDN switching and signaling standards.

For more information, see Chapter 11 of *CCNA Routing and Switching Exam Cram, Second Edition.*

Question 25

The correct answers are b and c. A data-link address is also known as a physical address, hardware address, or more commonly, a MAC address. (So named a MAC address because this address resides at the Media Access Control sublayer of the data-link layer.) The MAC address is the address "burned" into every network adapter card by the manufacturer. This address is considered a "flat" address, because no logical arrangement of these addresses is on a network. A flat addressing scheme simply gives each member a unique identifier that is associated with them. Answer b is incorrect, because data link addresses are considered flat addresses. Answer d is incorrect, because network addresses are considered hierarchical.

A network address, or logical address, is an address that resides at the network layer of the OSI model. Network addresses are hierarchical in nature. A hierarchical addressing scheme uses logically structured addresses to provide a more organized environment. Examples of hierarchical network addresses are IP and IPX. An IP address comprises a network ID, which identifies the network a host belongs to, and a host ID, which is unique to that host. An IPX address utilizes a similar network ID and host ID format in addressing. The hierarchical addressing of IP and IPX enable complex networks to be logically grouped and organized.

For more information, see Chapter 4 of *CCNA Routing and Switching Exam Cram, Second Edition.*

Question 26

The correct answer is d. Typing "**TERMINAL EDITING**" will enable the command history feature if it has been disabled; it is enabled by default. All of the other choices are not valid Cisco IOS commands; therefore, answers a, b, and d are incorrect. The Cisco IOS includes a command history feature that can be used to store previously entered commands in a history buffer. Commands in the history buffer can be recalled by an operator at the router prompt, using Ctrl+P and Ctrl+N, saving configuration time. The command history feature is enabled by default, but it can be disabled, and the size of the history buffer can be modified. The following list details the common commands used when modifying the command history:

➤ *SHOW HISTORY*—Typing the **SHOW HISTORY** command will display all of the commands currently stored in the history buffer. Pressing Ctrl+P will recall previous commands in the command history at the router prompt. (The up arrow will do this as well.) Pressing Ctrl+N will return to more recent commands in the command history after recalling previous commands. (The down arrow will do this as well.)

➤ *TERMINAL HISTORY SIZE (0-256)*—Typing "TERMINAL HIS-TORY SIZE (0-256)" will allow you to set the number of command lines the command history buffer will hold. You must enter a number between 0 and 256. (The default is 10 commands.)

> *TERMINAL NO EDITING*—Typing "**TERMINAL NO EDITING**" will disable the command history feature; it is enabled by default.

> *TERMINAL EDITING*—Typing "**TERMINAL EDITING**" will enable the command history feature if it has been disabled.

For more information, see Chapter 8 of *CCNA Routing and Switching Exam Cram, Second Edition.*

Question 27

The correct answer is a. Both Primary and Basic Rate ISDN utilize a D channel for control and signaling purposes. Answers b, c, and d are incorrect, because the D channel is not used to send data of any kind; it is just used for control and signaling purposes. ISDN is a communications standard that uses digital telephone lines to transmit voice, data, and video and is a dial-up service that is used on demand. The bandwidth provided by ISDN can be divided into two categories: Primary Rate ISDN (PRI) and Basic Rate ISDN (BRI). BRI (also known as 2B+1D) comprises two B channels that are 64Kbps each and one 16Kbps D channel. It uses the two B channels to transmit data and uses the 16Kbps D channel for control and signaling purposes. PRI (also known as 23B+1D) comprises 23 B channels that are 64KBps each and one 64Kbps D channel.

For more information, see Chapter 15 of *CCNA Routing and Switching Exam Cram, Second Edition.*

Question 28

The correct answer is d. All of the answers are valid entries for the protocol field in an extended IP access list except for **TELNET**; therefore, answers a, b, c, and e are incorrect. Telnet is used in the port field of an extended IP access list. The syntax for creating an extended IP access list is as follows:

```
ACCESS-LIST [access-list-number] [permit|deny] [protocol] [source]
 [wildcard mask] [destination] [wildmask] [operator][port]
```

> *ACCESS-LIST*—This is the command to denote an access list.

> *[access-list-number]*—This can be a number between 100 and 199 that is used to denote an extended IP access list.

> *[permit|deny]*—This will allow (permit) or disallow (deny) traffic specified in this access list.

➤ *[protocol]*—This parameter is used to specify the protocol that will be filtered in this access list. It can be a protocol name or number between 0 and 255. Common values are TCP (port 6), UDP (port 17), and IP (use of IP will denote all IP protocols).

➤ *[source]*—This specifies the source IP addressing information.

➤ *[wildcard mask]*—This parameter can be optionally applied to further define the source. A wildmask can be used to control access to an entire IP network ID rather than a single IP address. A wildcard mask will use the number 255 to mean "any" and the number 0 to mean "must match exactly." The terms *host* and *any* may also be used here to more quickly specify a single host or an entire network. The **HOST** keyword is the same as the wildcard mask 0.0.0.0. The **ANY** keyword is the same as the wildcard mask 255.255.255.255.

➤ *[destination]*—This specifies the destination IP addressing information.

➤ *[wildmask]*—This parameter can be optionally applied to further define the destination IP addressing.

➤ *[operator]*—This can be optionally applied to define how to interpret the value entered in the **[port]** section of the access list. Common values are equal to the port specified (**EQ**), less than the port number specified (**LT**), and greater than the port number specified (**GT**).

➤ *[port]*—This specifies the port number this access list will act on. Common port numbers include 21 (FTP), 23 (TELNET), 25 (SMTP), 53 (DNS), 69 (TFTP), and 80 (HTTP).

For more information, see Chapter 13 of *CCNA Routing and Switching Exam Cram, Second Edition*.

Question 29

The correct answers are a and b. X.25 is a packet switching technology that also utilizes logical or virtual circuits to facilitate WAN connectivity. The two types of logical circuits used in X.25 are SVC and PVC. An SVC is a temporary connection that is established on an as-needed basis and is closed when not in use.

A PVC is a permanent connection that is always available for use. Answer c is incorrect, because Local Management Interface (LMI) is a signaling format used by frame relay. Answer d is incorrect, because data-link connection identifier (DLCI) is a unique identifier used to identify a specific frame relay connection. Packet switching is a WAN technology that sends data over WAN links through the use of logical circuits, where data is encapsulated into packets and routed by the service provider through various switching points. The two most common packet switching technologies in use are frame relay and X.25. The following details the features of X.25 and frame relay:

Frame relay is a packet switching technology that utilizes logical circuits to form a logical connection, which is identified by the DLCI. The DLCI is a unique identifier used to identify a specific frame relay connection that is managed through a signaling format known as LMI, providing information about the DLCI values, as well as other connection-related information. Your Cisco router must utilize the same LMI signaling format as your service provider. LMI has three signaling formats: CISCO, ANSI, and Q933A. The Cisco default LMI type is CISCO.

For more information, see Chapter 16 of *CCNA Routing and Switching Exam Cram, Second Edition.*

Question 30

The correct answer is b. The **CLEAR ACCESS-LIST COUNTERS** command will clear the previous matches for the lines in an access list. Answer a is incorrect, because **LINES** is not a valid Cisco IOS command. Answer c is incorrect, because the **CLEAR ACCESS-LIST COUNTERS** command contains hyphens between **ACCESS** and **LIST**. Answer d is incorrect, because the **CLEAR ACCESS-LIST COUNTERS** command does not use a hyphen between **LIST** and **COUNTERS**. The **SHOW ACCESS-LIST** command will display all access lists in use on the router and will also show each line of the access list and return the number of times a packet matched that line. To clear the matches, use the **CLEAR ACCESS-LIST COUNTERS** command. The syntax for this command is:

```
CLEAR ACCESS-LIST COUNTERS [access list number]
```

For more information, see Chapter 13 of *CCNA Routing and Switching Exam Cram, Second Edition.*

Question 31

The correct answer is a. The access list **ACCESS-LIST 100 PERMIT TCP HOST 200.200.200.50 131.200.0.0 0.0 255.255 EQ 23** will allow host **200.200.200.50** to have Telnet access to network **131.200.0.0**. Answer b is incorrect, because the wildcard mask 255.255.255.0 will allow **TELNET** access to all IP addresses that have 50 in the last octet. Do not confuse a wildcard mask with a subnet mask! A wildcard mask will use the number 255 to mean "any" and the number 0 to mean "must match exactly." Answer c is incorrect, because the **ACCESS-LIST** command does not begin with IP. Answer d is incorrect, because the access list number is 10, which is used for standard access lists, not for extended lists. Extended access lists use numbers between 100 and 199. The syntax for creating an extended IP access list is as follows:

```
ACCESS-LIST [access-list-number] [permit|deny] [protocol] [source]
 [wildcard mask] [destination] [wildmask] [operator][port]
```

➤ *ACCESS-LIST*—This is the command to denote an access list.

➤ *[access-list-number]*—This can be a number between 100 and 199 and is used to denote an extended IP access list.

➤ *[permit|deny]*—This will allow (permit) or disallow (deny) traffic specified in this access list.

➤ *[protocol]*—This is used to specify the protocol that will be filtered in this access list. Common values are TCP, UDP, and IP (use of IP will denote all IP protocols).

➤ *[source]*—This specifies the source IP addressing information.

➤ *[wildcard mask]*—This can be optionally applied to further define the source. A wildmask can be used to control access to an entire IP network ID rather than a single IP address. A wildcard mask will use the number 255 to mean "any" and the number 0 to mean "must match exactly." The terms *host* and *any* may also be used here to more quickly specify a single host or an entire network. The **HOST** keyword is the same as the wildcard mask 0.0.0.0. The **ANY** keyword is the same as the wildcard mask 255.255.255.255.

➤ *[destination]*—This specifies the destination IP addressing information.

➤ *[wildmask]*—This can be optionally applied to further define the destination IP addressing.

➤ *[operator]*—This can be optionally applied to define how to interpret the value entered in the **[port]** section of the access list. Common values are equal to the port specified (**EQ**), less than the port number specified (**LT**), and greater than the port number specified (**GT**).

➤ *[port]*—This specifies the port number this access list will act on. Common port numbers include 21 (FTP), 23 (TELNET), 25 (SMTP), 53 (DNS), 69 (TFTP), and 80 (HTTP).

For more information, see Chapter 13 of *CCNA Routing and Switching Exam Cram, Second Edition.*

Question 32

The correct answer is a. A switch is a data-link layer connectivity device used to segment a network. A switch resides at the data-link layer of the OSI model and can filter traffic based on the data link or MAC address of a packet. (The MAC address is the address burned into the network adapter card by the manufacturer.)

Cisco switches employ two basic methods of forwarding packets: cut-through and store-and-forward. Store-and-forward switches will copy an incoming packet to the local buffer and perform error checking on the packet before sending it on to its destination. If the packet contains an error, it is discarded. Cut-through switches will read the destination address of an incoming packet and immediately search its switch table for a destination port. Cut-through switches will not perform any error checking of the packet. Answers b and c are incorrect, because of the fact that cut-through switches do not perform error checking or copy the packet to their buffer before processing, they experience less latency (delay) than store-and-forward switches. Answer d is incorrect, because there is no such thing as cut-and-forward switching.

For more information, see Chapter 6 of *CCNA Routing and Switching Exam Cram, Second Edition*.

Question 33

The correct answers are a, b, and e. The Cisco IOS is loaded into memory per instructions from the boot system command. The IOS can be loaded from Flash, ROM, or a network location (TFTP server). Answer c is incorrect, because no information is permanently stored in RAM. Answer f is incorrect, because Cisco routers do not utilize floppy disk drives. The Cisco router startup sequence is divided into three basic steps:

1. The bootstrap program in ROM executes the POST, which performs a basic hardware level check.

2. The Cisco IOS is loaded into memory per instructions from the boot system command. The IOS can be loaded from Flash, ROM, or a network location (TFTP server).

3. The router's configuration file is loaded into memory from NVRAM. If no configuration file is found, the setup program automatically initiates the dialog necessary to create a configuration file.

For more information, see Chapter 4 of *CCNA Routing and Switching Exam Cram, Second Edition*.

Question 34

The correct answer is b. The OSI model defines standards used in networking and comprises a seven-layer model. The presentation layer of the OSI model is primarily used as a "translator" for the application layer and uses Abstract Syntax Notation One (ASN 1) to negotiate or to translate information to the application layer. The following list outlines the seven layers of the OSI model and their functions:

➤ *Application*—The "window" to networking used by programs, the application layer is responsible for:

➤ Verifying that the appropriate resources are present to initiate a connection with the destination node.

➤ Verifying the identity of the destination node.

Application layer protocols include Telnet, File Transfer Protocol (FTP), Simple Mail Transfer Protocol (SMTP), World Wide Web (WWW), Electronic Data Interchange (EDI), and Wide Area Information Server (WAIS).

➤ *Presentation*—Essentially a translator, the presentation layer is responsible for:

➤ Translating text and data syntax, such as Extended Binary-Coded Decimal Interchange Code (EBCDIC, used in IBM systems) to American Standard Code for Information Interchange (ASCII, used in PC and most computer systems).

➤ Using Abstract Syntax Notation to perform data translation.

Presentation layer protocols include PICTure (PICT, Apple computer picture format), Tagged Image File Format (TIFF), Joint Photographic Experts Group (JPEG), Musical Instrument Digital Interface (MIDI), Motion Picture Expert Group (MPEG), and Quick Time (audio/video application).

➤ *Session*—This is used to coordinate communication between nodes; the session layer is responsible for:

➤ Creating, maintaining, and ending a communication session.

➤ Coordinating service requests and responses that occur between nodes across the network.

Session layer protocols include Network File System (NFS, used by SUN Microsystems and Unix with TCP/IP), structured query language (SQL, used to define database information requests), Remote Procedure Calls (RPC, used in Microsoft network communication), X Window (used by Unix terminals),

AppleTalk Session Protocol (ASP, used by Apple computers), and Digital Network Architecture Session Control Protocol (DNA SCP, used by IBM).

➤ *Transport*—Reliable end-to-end communication is the primary function of the transport layer. Its many responsibilities include:

➤ Ensuring flow control so the amount of data being sent from one node will not overwhelm the destination node.

➤ Ensuring the ability of multiple applications to utilize a single transport (multiplexing).

➤ Ensuring the reliable transfer of data. The transport layer implements connection-oriented services between nodes, which utilizes a three-way handshake (synchronization, acknowledgment, and data transfer) to efficiently transfer data.

➤ Ensuring positive acknowledgment, which is the process of a node waiting for an acknowledgment from the destination node prior to sending data.

➤ Windowing, which is a form of flow control that specifies how much data will be transferred between acknowledgments.

Transport layer protocols include Transmission Control Protocol (TCP, part of the TCP/IP protocol suite), and Sequenced Packet Exchange (SPX, part of the IPX/SPX protocol suite).

➤ *Network*—This selects the appropriate path a packet should take to get to its intended destination. It is responsible for routing, which is the process of using a network layer address to determine the best path a packet will travel to its destination. Network layer protocols include Internet Protocol (IP, part of the TCP/IP protocol suite), and Internetwork Packet Exchange (IPX, part of the IPX/SPX protocol suite).

➤ *Data Link*—Divided into two sublayers (Media Access Control [MAC], and Logical Link Control [LLC]), the data-link layer is responsible for:

➤ Preparing data from upper layers to be transmitted over the physical medium by encapsulating upper-layer data into frames. This frame includes the source and destination of MAC addresses in the frame header.

➤ Converting data into bits, so it can be transmitted by the physical layer.

➤   Adding a cyclical redundancy check (CRC) to the end of a frame, used for error checking at the data-link layer.

Data-link protocols include High-Level Data Link Control (HDLC, the Cisco default encapsulation for serial connections), Synchronous Data Link Control, (SDLC, used in IBM networks), Link Access Procedure Balanced (LAPB, used with X.25), X.25 (packet switching network), Serial Line Internet Protocol (SLIP, an older TCP/IP dial-up protocol), Point-to-Point Protocol (PPP, a newer dial-up protocol), Integrated Services Digital Network (ISDN, a dial-up digital service), and frame relay (a packet switching network).

➤   *Physical*—The lowest layer of the OSI model, the physical layer defines the electrical functionality required to send and receive bits over a given physical medium. Specifications that define the voltage levels and physical components of a network are defined at the physical layer. Protocols are not specified at the physical layer, because they are implemented as software. Examples of the standards for sending data over the physical medium are Ethernet (the most widely used standard in networking), Token Ring (IBM's proprietary networking topology), and Fiber Distributed Data Interface (FDDI, a standard for fiber optic networks, commonly used as a backbone).

For more information, see Chapter 4 of *CCNA Routing and Switching Exam Cram, Second Edition*.

Question 35

The correct answer is b. The **SHOW IPX INTERFACE S0** command will display a list of all the IP access lists that are currently in use on interface **S0** of your Cisco router. Answer a is incorrect, because the **SHOW IPX ACCESS-LISTS** command cannot be followed with an interface number. Answer c is incorrect, because **SHOW IPX ACCESS-LISTS** will not display what interface an access list has been applied to, as the **SHOW INTERFACE** commands do. Answer d is incorrect, because **SHOW IPX INTERFACE-ACCESS-LISTS S0** is not a valid Cisco IOS command. Access lists can be used to filter the traffic that is handled by a Cisco router. Several Cisco IOS commands can be used to view the various access lists in use on a Cisco router. The following list details the most common Cisco IOS commands used to view access lists:

➤   *SHOW ACCESS-LISTS*—This command will display all of the access lists in use on the router. It will also show each line of the access list and return the number of times a packet matched that line. This command can also be used with a specific access list number to display this detail on a single IP access list. **SHOW ACCESS-LISTS** will not display what interface an access list has been applied to, as the **SHOW INTER-FACE** commands do.

➤ *SHOW IP ACCESS-LIST*—This command will display all IP access lists in use on the router. It will also show each line of the access list and return the number of times a packet matched that line. This command can also be used with a specific IP access list number to display this detail on a single IP access list. **SHOW IP ACCESS-LISTS** will not display what interface an access list has been applied to, as the **SHOW INTERFACE** commands do.

➤ *SHOW IPX ACCESS-LIST*—This command will display all IP access lists in use on the router. It will also show each line of the access list and return the number of times a packet matched that line. This command can also be used with a specific IPX access list number to display this detail on a single IPX access list. **SHOW IPX ACCESS-LISTS** will not display what interface an access list has been applied to, as the **SHOW INTERFACE** commands do.

➤ *SHOW IP INTERFACE [interface number]*—This will display the IP access lists that have been applied to a specific interface. It will not display any matches to the individual lines of the access list, as the **SHOW ACCESS-LIST** commands do.

➤ *SHOW IPX INTERFACE [interface number]*—This will display the IP access lists that have been applied to a specific interface. It will not display any matches to the individual lines of the access list like the **SHOW ACCESS-LIST** commands.

For more information, see Chapter 8 of *CCNA Routing and Switching Exam Cram, Second Edition.*

Question 36

The correct answer is a. The data-link layer of the OSI model is divided into two sublayers, MAC and LLC, and is responsible for:

➤ Preparing data from upper layers to be transmitted over the physical medium by encapsulating upper-layer data into frames. This frame includes the source and destination of MAC addresses in the frame header.

➤ Converting data into bits, so it can be transmitted by the physical layer.

➤ Adding a Cyclical Redundancy Check (CRC) to the end of a frame that is used for error checking at the data-link layer.

Data-link protocols include HDLC (the Cisco default encapsulation for serial connections), SDLC (used in IBM networks), LAPB (used with X.25), X.25 (a Packet switching network), SLIP (an older TCP/IP dial-up protocol), PPP (a newer dial-up protocol), ISDN (a dial-up digital service), and frame relay (a packet switching network).

RPC is a session layer protocol.

For more information, see Chapter 4 of *CCNA Routing and Switching Exam Cram, Second Edition.*

Question 37

The correct answer is f. The OSI model defines standards used in networking and comprises a seven-layer model. The data-link layer of the OSI model is responsible for preparing data from upper layers to be transmitted over the physical medium by encapsulating upper-layer data into frames and converts data into bits for transmission by the physical layer. Since no other layer of the OSI model is responsible for preparing data from upper layers to be transmitted over the physical medium, all of the other answers are incorrect. The following list outlines the seven layers of the OSI model and their functions:

➤ *Application*—The "window" to networking used by programs, the application layer is responsible for:

  ➤ Verifying that the appropriate resources are present to initiate a connection with the destination node.

  ➤ Verifying the identity of the destination node.

Application layer protocols include Telnet, File Transfer Protocol (FTP), Simple Mail Transfer Protocol (SMTP), World Wide Web (WWW), Electronic Data Interchange (EDI), and Wide Area Information Server (WAIS).

➤ *Presentation*—Essentially a translator, the presentation layer is responsible for:

  ➤ Translating text and data syntax, such as Extended Binary-Coded Decimal Interchange Code (EBCDIC, used in IBM systems) to American Standard Code for Information Interchange (ASCII, used in PC and most computer systems).

  ➤ Using Abstract Syntax Notation to perform data translation.

Presentation layer protocols include PICTure (PICT, Apple computer picture format), Tagged Image File Format (TIFF), Joint Photographic Experts Group (JPEG), Musical Instrument Digital Interface (MIDI), Motion Picture Expert Group (MPEG), and Quick Time (audio/video application).

➤ *Session*—This is used to coordinate communication between nodes; the session layer is responsible for:

> ➤ Creating, maintaining, and ending a communication session.

> ➤ Coordinating service requests and responses that occur between nodes across the network.

Session layer protocols include Network File System (NFS, used by SUN Microsystems and Unix with TCP/IP), structured query language (SQL, used to define database information requests), Remote Procedure Calls (RPC, used in Microsoft network communication), X Window (used by Unix terminals), AppleTalk Session Protocol (ASP, used by Apple computers), and Digital Network Architecture Session Control Protocol (DNA SCP, used by IBM).

➤ *Transport*—Reliable end-to-end communication is the primary function of the transport layer. Its many responsibilities include:

> ➤ Ensuring flow control so the amount of data being sent from one node will not overwhelm the destination node.

> ➤ Ensuring the ability of multiple applications to utilize a single transport (multiplexing).

> ➤ Ensuring the reliable transfer of data. The transport layer implements connection-oriented services between nodes, which utilizes a three-way handshake (synchronization, acknowledgment, and data transfer) to efficiently transfer data.

> ➤ Ensuring positive acknowledgment, which is the process of a node waiting for an acknowledgment from the destination node prior to sending data.

> ➤ Windowing, which is a form of flow control that specifies how much data will be transferred between acknowledgments.

Transport layer protocols include Transmission Control Protocol (TCP, part of the TCP/IP protocol suite), and Sequenced Packet Exchange (SPX, part of the IPX/SPX protocol suite).

➤ *Network*—This selects the appropriate path a packet should take to get to its intended destination. It is responsible for routing, which is the process of using a network layer address to determine the best path a packet will travel to its destination. Network layer protocols include Internet Protocol (IP, part of the TCP/IP protocol suite), and Internetwork Packet Exchange (IPX, part of the IPX/SPX protocol suite).

➤ *Data Link*—Divided into two sublayers (Media Access Control [MAC], and Logical Link Control [LLC]), the data-link layer is responsible for:

➤ Preparing data from upper layers to be transmitted over the physical medium by encapsulating upper-layer data into frames. This frame includes the source and destination of MAC addresses in the frame header.

➤ Converting data into bits, so it can be transmitted by the physical layer.

➤ Adding a cyclical redundancy check (CRC) to the end of a frame, used for error checking at the data-link layer.

Data-link protocols include High-Level Data Link Control (HDLC, the Cisco default encapsulation for serial connections), Synchronous Data Link Control, (SDLC, used in IBM networks), Link Access Procedure Balanced (LAPB, used with X.25), X.25 (packet switching network), Serial Line Internet Protocol (SLIP, an older TCP/IP dial-up protocol), Point-to-Point Protocol (PPP, a newer dial-up protocol), Integrated Services Digital Network (ISDN, a dial-up digital service), and frame relay (a packet switching network).

➤ *Physical*—The lowest layer of the OSI model, the physical layer defines the electrical functionality required to send and receive bits over a given physical medium. Specifications that define the voltage levels and physical components of a network are defined at the physical layer. Protocols are not specified at the physical layer, because they are implemented as software. Examples of the standards for sending data over the physical medium are Ethernet (the most widely used standard in networking), Token Ring (IBM's proprietary networking topology), and Fiber Distributed Data Interface (FDDI, a standard for fiber optic networks, commonly used as a backbone).

For more information, see Chapter 4 of *CCNA Routing and Switching Exam Cram, Second Edition*.

Question 38

The correct answer is a. The OSI model defines a layered approach to network communication. Each layer of the OSI model adds its own layer-specific information and passes it on to the next layer until it leaves the computer and goes out onto the network. This process is known as encapsulation, and it involves five basic steps:

1. User information is converted to data.

2. Data is converted to segments.

3. Segments are converted to packets or datagrams.

4. Packets and datagrams are converted to frames.

5. Frames are converted to bits.

Answer b is incorrect, because frames are converted to bits in the fifth step of encapsulation. Answer c is incorrect, because packets and datagrams are converted to frames in the fourth step of encapsulation. Answer d is incorrect, because segments are converted to packets, not to data.

For more information, see Chapter 3 of *CCNA Routing and Switching Exam Cram, Second Edition.*

Question 39

The correct answer is b. FTP, Telnet, EDI, and WAIS are all application layer protocols. Answer a is incorrect, because SMTP and FTP are application layer protocols, and MIDI and JPEG are presentation layer protocols. Answer c is incorrect, because Telnet is an application layer protocol, and WAIS, MIDI, and MPEG are presentation layer protocols. Answer d is incorrect, because WAIS is an application layer protocol and PICT, MIDI, and Quick Time are presentation layer protocols. The OSI model defines standards used in networking and comprises a seven-layer model. The presentation layer functions as a translator and is responsible for text and data syntax translation, such as Extended Binary-Coded Decimal Interchange Code (EBCDIC, used in IBM systems) to American Standard Code for Information Interchange (ASCII, used in PC and most computer systems). Presentation layer protocols include PICT, TIFF, JPEG, MIDI, MPEG, and Quick Time. The application layer of the OSI model functions as the "window" for applications to access network resources and is responsible for verifying the identity of the destination node. Application layer protocols include Telnet, FTP, SMTP, WWW, EDI, and WAIS.

For more information, see Chapter 3 of *CCNA Routing and Switching Exam Cram, Second Edition.*

Question 40

The correct answer is d. The OSI model defines standards used in networking and comprises a seven-layer model. The transport layer of the OSI model is responsible for flow control, multiplexing, and end-to-end communication.

Since no other layer of the OSI model is responsible for flow control, multiplexing, and end-to-end communication, all of the other answers are incorrect. The following list outlines the seven layers of the OSI model and their functions:

➤ *Application*—The "window" to networking used by programs, the application layer is responsible for:

    ➤ Verifying that the appropriate resources are present to initiate a connection with the destination node.

    ➤ Verifying the identity of the destination node.

Application layer protocols include Telnet, File Transfer Protocol (FTP), Simple Mail Transfer Protocol (SMTP), World Wide Web (WWW), Electronic Data Interchange (EDI), and Wide Area Information Server (WAIS).

➤ *Presentation*—Essentially a translator, the presentation layer is responsible for:

    ➤ Translating text and data syntax, such as Extended Binary-Coded Decimal Interchange Code (EBCDIC, used in IBM systems) to American Standard Code for Information Interchange (ASCII, used in PC and most computer systems).

    ➤ Using Abstract Syntax Notation to perform data translation.

Presentation layer protocols include PICTure (PICT, Apple computer picture format), Tagged Image File Format (TIFF), Joint Photographic Experts Group (JPEG), Musical Instrument Digital Interface (MIDI), Motion Picture Expert Group (MPEG) and Quick Time (audio/video application).

➤ *Session*—This is used to coordinate communication between nodes; the session layer is responsible for:

    ➤ Creating, maintaining, and ending a communication session.

    ➤ Coordinating service requests and responses that occur between nodes across the network.

Session layer protocols include Network File System (NFS, used by SUN Microsystems and Unix with TCP/IP), structured query language (SQL, used to define database information requests), Remote Procedure Calls (RPC, used in Microsoft network communication), X Window (used by Unix terminals), AppleTalk Session Protocol (ASP, used by Apple computers) and Digital Network Architecture Session Control Protocol (DNA SCP, used by IBM).

➤ *Transport*—Reliable end-to-end communication is the primary function of the transport layer. Its many responsibilities include:

➤ Ensuring flow control so the amount of data being sent from one node will not overwhelm the destination node.

➤ Ensuring the ability of multiple applications to utilize a single transport (multiplexing).

➤ Ensuring the reliable transfer of data. The transport layer implements connection-oriented services between nodes, which utilizes a three-way handshake (synchronization, acknowledgment, and data transfer) to efficiently transfer data.

➤ Ensuring positive acknowledgment, which is the process of a node waiting for an acknowledgment from the destination node prior to sending data.

➤ Windowing, which is a form of flow control that specifies how much data will be transferred between acknowledgments.

Transport layer protocols include Transmission Control Protocol (TCP, part of the TCP/IP protocol suite), and Sequenced Packet Exchange (SPX, part of the IPX/SPX protocol suite).

➤ *Network*—This selects the appropriate path a packet should take to get to its intended destination. It is responsible for routing, which is the process of using a network layer address to determine the best path a packet will travel to its destination. Network layer protocols include Internet Protocol (IP, part of the TCP/IP protocol suite) and Internetwork Packet Exchange (IPX, part of the IPX/SPX protocol suite).

➤ *Data Link*—Divided into two sublayers (Media Access Control [MAC] and Logical Link Control [LLC]), the data-link layer is responsible for:

➤ Preparing data from upper layers to be transmitted over the physical medium by encapsulating upper-layer data into frames. This frame includes the source and destination of MAC addresses in the frame header.

➤ Converting data into bits, so it can be transmitted by the physical layer.

➤ Adding a cyclical redundancy check (CRC) to the end of a frame, used for error checking at the data-link layer.

Data-link protocols include High-Level Data Link Control (HDLC, the Cisco default encapsulation for serial connections), Synchronous Data Link Control, (SDLC, used in IBM networks), Link Access Procedure Balanced (LAPB, used with X.25), X.25 (packet switching network), Serial Line Internet Protocol (SLIP, an older TCP/IP dial-up protocol), Point-to-Point Protocol (PPP, a newer dial-up protocol), Integrated Services Digital Network (ISDN, a dial-up digital service) and frame relay (a packet switching network).

➤ *Physical*—The lowest layer of the OSI model, the physical layer defines the electrical functionality required to send and receive bits over a given physical medium. Specifications that define the voltage levels and physical components of a network are defined at the physical layer. Protocols are not specified at the physical layer, because they are implemented as software. Examples of the standards for sending data over the physical medium are Ethernet (the most widely used standard in networking), Token Ring (IBM's proprietary networking topology) and Fiber Distributed Data Interface (FDDI, a standard for fiber optic networks, commonly used as a backbone).

For more information, see Chapter 4 of *CCNA Routing and Switching Exam Cram, Second Edition.*

Question 41

The correct answer is b. The access list **ACCESS LIST 922 DENY −1 22 0 37 0** will deny IPX network 22 access to IPX network 37. Answer a is incorrect, because this access list only denies access to IPX network 22. The "-1" in an extended IPX access list in the protocol field means "all protocols," not "all networks." Answer c is incorrect, because the source network is always specified first in an extended IPX access list. Answer d is incorrect, because only IPX network 22 is denied access to IPX network 37. The access list **ACCESS LIST 922 PERMIT −1 -1 0 -1 0** will allow all other networks access to IPX network 37.

An extended IPX access list can be used to filter traffic based on source and destination IPX address, as well as IPX protocol and IPX socket number. The syntax for creating an extended IPX access list is:

```
ACCESS-LIST [access-list-number] [permit|deny] [protocol] [source]
 [source-node-mask] [source socket] [destination]
 [[destination-node-mask] [destination socket]
```

➤ *ACCESS-LIST*—This command is followed by the **access list number**. When using a standard IPX access list, this can be a number between 800 and 899.

➤ *[permit|deny]*—This statement will either allow or disallow traffic in this access list.

➤ *[source]*—This statement is used to specify the source IPX addressing information. A "-1" can be used to signify "all networks."

➤ *[destination]*—This statement is used to specify destination addressing information ("-1" may be used here as well).

➤ *[source-node-mask]* and *[destination-node-mask]*—These are optional parameters for controlling access to a portion of a network, similar to the wildmask that is used in IP access lists.

➤ *[protocol]*—This statement is used to specify the IPX protocol to filter. "-1" can be used to specify "all protocols."

➤ *[source socket]* and *[destination socket]*—This statement is used to specify the IPX socket number to filter. "0" can be used to specify "all sockets."

For more information, see Chapter 13 of *CCNA Routing and Switching Exam Cram, Second Edition.*

Question 42

The correct answer is b. The **SHOW IP ROUTE** command is used to display the contents of the routing table and will display all known routes on the entire router. Answer a is incorrect, because **SHOW ROUTING TABLE** is not a valid Cisco IOS command. Answer c is incorrect, because the **SHOW IP ROUTE** command does not have a hyphen between **IP** and **ROUTE**. Answer d is incorrect, because **SHOW IP PROTOCOL** will display the routing protocols in use for IP on the entire router, as well as update frequency and filter information, but will not display the routing table.

For more information, see Chapter 11 of *CCNA Routing and Switching Exam Cram, Second Edition.*

Question 43

The correct answers are a, b, and c. Flow control is a method used by TCP at the transport layer that ensures the amount of data being sent from one node will not overwhelm the destination node. Three basic techniques are used in flow control:

➤ *Windowing*—This is the process of predetermining the amount of data that will be sent before an acknowledgment is expected.

➤ *Buffering*—This is a method of temporarily storing packets in memory buffers until they can be processed by the computer.

➤ *Source-Quench Messaging*—This is a message sent by a receiving computer when its memory buffers are nearing capacity. A source-quench message is a device's way of telling the sending computer to slow down the rate of transmission.

Answer d is incorrect, because multiplexing is the ability of multiple connections to utilize a single transport and is not used as part of flow control.

For more information, see Chapter 9 of *CCNA Routing and Switching Exam Cram, Second Edition.*

Question 44

The correct answer is b. To enter privileged mode, enter the **ENABLE** command from user mode. Answer a is incorrect, because the **CONFIGURE** command cannot be executed in user mode. Answers c and d are incorrect, because there are no such modes as system and router. The Cisco IOS uses a command interpreter known as EXEC, which contains two primary modes: user mode and privileged mode. User mode is used to display basic router system information and to connect to remote devices. The commands in user mode are limited in capability. When you first log into a router, you are placed in user mode, where the router prompt is followed by an angle bracket (**ROUTER>**). Privileged mode is used to modify and view the configuration of the router, as well as set IOS parameters. In this mode, the router prompt is followed by a pound sign (**ROUTER#**). To enter privileged mode, enter the **ENABLE** command from user mode.

For more information, see Chapter 8 of *CCNA Routing and Switching Exam Cram, Second Edition.*

Question 45

The correct answer is b. The range of numbers used to specify an extended IP access list is 100 through 199. Answer a is incorrect, because an IP standard access list uses the range 1–99. Answer c is incorrect, because an IPX standard access list uses the range 800–899. Answer d is incorrect, because an IPX extended access list uses the range 900–999. Access lists can be easily identified by referring to the number after the **ACCESS-LIST** command. The five major types of access lists utilize the following ranges of numbers:

➤ *IP Standard Access List—1-99*

➤ *IP Extended Access List—100-199*

➤ *IPX Standard Access List—800-899*

➤ *IPX Extended Access List—900-999*

➤ *IPX SAP Access List—1000-1099*

For more information, see Chapter 13 of *CCNA Routing and Switching Exam Cram, Second Edition.*

Question 46

The correct answer is b. The local loop is the cabling that extends from the DEMARC to the telephone company's central office (CO). Answer a is incorrect, because the CPE devices are the communication devices (telephones, modems, and so forth) that exist at the customer's location. Answer c is incorrect, because the local loop does not extend from an ISP to the telephone company. Answer d is incorrect, because the DEMARC is the point at the customer's premises where CPE devices connect. When WANs communicate over telecommunication lines, several components are used to facilitate the connection. The basic process of WAN communication begins with a call placed using a CPE, or Customer Premises Equipment. The CPE connects to the DEMARC, which passes the data through the local loop to the CO, which will act as the central distribution point for sending and receiving information. The following list details the common components of WAN communication:

➤ *Customer Premises Equipment (CPE)*—These devices are the communication devices (telephones, modems, and so forth) that exist at the customer's location.

➤ *Demarcation (DEMARC)*—This is the point at the customer's premises where CPE devices connect. The telephone company owns the DEMARC, which is usually a large punch-down board located in a wiring closet.

➤ *Local Loop*—This is the cabling that extends from the DEMARC to the telephone company's CO.

➤ *Central Office (CO)*—This is the telephone company's office that acts as the central communication point for the customer.

For more information, see Chapter 14 of *CCNA Routing and Switching Exam Cram, Second Edition.*

Question 47

The correct answers are a, b, and c. Transmission Control Protocol (TCP) is a transport layer protocol that is connection-oriented. Answer d is incorrect, because connection-oriented protocols guarantee packet delivery. Connection-oriented protocols feature:

➤ *Reliability*—A communication session is established between hosts before sending data. This session is considered a virtual circuit.

➤ *Sequencing*—When a connection-oriented protocol like TCP sends data, it numbers each segment so that the destination host can receive the data in the proper order.

➤ *Guaranteed Delivery*—Connection-oriented protocols utilize error checking to guarantee packet delivery.

➤ *More Overhead*—Because of the additional error checking and sequencing responsibilities, connection-oriented protocols require more overhead than connectionless protocols.

➤ *User Datagram Protocol (UDP)*—This is a transport layer protocol that is connectionless. Connectionless protocols feature:

➤ No sequencing or virtual circuit creation. Connectionless protocols do not sequence packets or create virtual circuits.

➤ No guarantee of delivery. Packets are sent as datagrams, and delivery is not guaranteed by connectionless protocols. When using a connectionless protocol like UDP, the guarantee of delivery is the responsibility of higher layer protocols.

➤ Less overhead. Because connectionless protocols do not perform any of the above services, overhead is less than connection-oriented protocols.

For more information, see Chapter 9 of *CCNA Routing and Switching Exam Cram, Second Edition.*

Question 48

The correct answer is b. NVRAM is used by Cisco routers to store the startup configuration. Answer a is incorrect, because ROM is a physical chip installed on a router's motherboard that contains the bootstrap program, the POST, and the operating system software (Cisco IOS). Answer c is incorrect, because the Flash is an erasable, programmable memory area that contains the Cisco operating system software (IOS). Answer d is incorrect, because RAM is used as the router's main working area and contains the running configuration.

The internal components of Cisco routers are comparable to that of conventional PCs. A Cisco router is similar to a PC in that it contains an operating system (known as the IOS), a bootstrap program (like the BIOS used in a PC), RAM (memory), and a processor (CPU). Unlike a PC, a Cisco router has no external drives to install software or upgrade the IOS. Rather than use a floppy or CD drive to enter software as PCs do, Cisco routers are updated by downloading update information from a server. The following list outlines the core components of Cisco routers:

➤ *Read-Only Memory (ROM)*—This is a physical chip installed on a router's motherboard that contains the bootstrap program, the POST, and the operating system software (Cisco IOS). When a Cisco router is first powered up, the bootstrap program and the POST are executed. ROM cannot be changed through software; the chip must be replaced if any modifications are needed.

➤ *Random Access Memory (RAM)*—As with PCs, RAM is used as the router's main working area and contains the running configuration. All information in RAM dissipates when the router is turned off.

➤ *Non-Volatile RAM (NVRAM)*—This is used by Cisco routers to store the startup or running configuration. After a configuration is created, it exists and runs in RAM. Because RAM cannot permanently store this information, NVRAM is used. Information in NVRAM is preserved after the router is powered off.

➤ *Flash*—Essentially the PROM used in PCs, the Flash is an erasable, programmable memory area that contains the Cisco operating system software (IOS). When a router's operating system needs to be upgraded, new software can be downloaded from a TFTP server to the router's Flash. Upgrading the IOS in this manner is typically more convenient than replacing the ROM chip in the router's motherboard. Information in Flash is retained when the router is powered off.

For more information, see Chapter 4 of *CCNA Routing and Switching Exam Cram, Second Edition*.

Question 49

The correct answer is b. A Cisco router can be configured to display a message when users log in. This message is known as a *logon banner* and can be configured using the **BANNER MOTD** command. Answer a is incorrect, because there is no such thing as medial optic transfer dilation, answer c is incorrect, because there is no such thing as routing protocol headers. Answer d is incorrect, because there is no such thing as distance vector headers. The **BANNER MOTD** command must be entered in global configuration mode. The correct syntax for setting a logon banner is as follows:

```
BANNER MOTD #     (note that the pound sign (#) is a delimiting
 character of your choice) [Message text] #
```

An example of a logon banner would be:

```
Router(config)#banner motd #
Enter TEXT message. End with the character '#'.
This is the message text of my logon banner #
Router(config)#^Z
```

For more information, see Chapter 8 of *CCNA Routing and Switching Exam Cram, Second Edition.*

Question 50

The correct answers are a and d. *Routing* refers to the process of determining the path a packet will take to its destination. Answer b is incorrect, because static routes are updated using the **IP ROUTE** command, not dynamic routing protocols. Answer c is incorrect, because dynamic routing tables generate more network traffic than static routing tables. A Cisco router, which resides at the network layer of the OSI model, uses a routing table of network layer addresses (IP addresses) to perform the routing of packets. A Cisco router cannot route anything unless there is an entry in its routing table that contains information on where to send the packet. A routing table can be of two basic types: static and dynamic. A static routing table is one that has its entries entered manually by an operator. On a Cisco router, a static route can be entered using the **IP ROUTE** command. A dynamic routing table is one that has its entries entered automatically by the router, through the use of a routing protocol such as RIP or IGRP. Static routing tables generate more network traffic than dynamic routing tables, because of the additional network traffic generated by the updating of dynamic routing protocols.

For more information, see Chapter 11 of *CCNA Routing and Switching Exam Cram, Second Edition.*

Question 51

The correct answer is b. **CONFIGURE TERMINAL** is used to enter configuration commands into the router from the console port or through Telnet. The **CONFIGURE** command, executed in privileged mode, is used to enter the configuration mode of a Cisco router. The four parameters used with the **CONFIGURE** command are **TERMINAL, MEMORY, OVERWRITE-NETWORK**, and **NETWORK**. The following list outlines the functions of these parameters:

➤ *CONFIGURE TERMINAL (CONFIG T)*—This command is used to enter configuration commands into the router from the console port or through Telnet.

➤ *CONFIGURE NETWORK (CONFIG NET)*—This command is used to copy the configuration file from a TFTP server into the router's RAM.

➤ *CONFIGURE OVERWRITE-NETWORK (CONFIG O)*—This command is used to copy a configuration file into NVRAM from a TFTP server. Use of the **CONFIGURE OVERWRITE-NETWORK** will not alter the running configuration.

➤ *CONFIGURE MEMORY (CONFIG MEM)*—This command is used to execute the configuration stored in NVRAM. It will copy the startup configuration to the running configuration.

For more information, see Chapter 8 of *CCNA Routing and Switching Exam Cram, Second Edition.*

Question 52

The correct answer is c. The **SETUP** command was used in this question to enter the System Configuration Dialog. The **SETUP** command will execute the System Configuration Dialog, which can be used to modify the configuration used by the router. The **SETUP** command must be run from privileged mode. Answer a is incorrect, because the **SHOW RUNNING-CONFIG** command can be used in privileged mode to display the current running configuration in RAM. Answers b and d are incorrect, because the **SHOW STARTUP-CONFIG** command can be used in privileged mode to display the startup configuration file stored in NVRAM.

For more information, see Chapter 5 of *CCNA Routing and Switching Exam Cram, Second Edition.*

Question 53

The correct answer is a. The **SHOW IP ACCESS LIST** command will display all IP access lists in use on the router. Answer b is incorrect, because **DISPLAY** is not a valid Cisco IOS command. Answer c is incorrect, because **ACCESS-LIST-IN-USE** is not a valid Cisco IOS command. Answer d is incorrect, because the **SHOW IP ACCESS-LIST** command cannot be followed with **INT E0**. Access lists can be used to filter the traffic that is handled by a Cisco router. Several Cisco IOS commands can be used to view the various access lists in use on a Cisco router. The following list details the most common Cisco IOS commands used to view access lists:

➤ *SHOW ACCESS-LISTS*—This command will display all of the access lists in use on the router. It will also show each line of the access list and return the number of times a packet matched that line. This command can also be used with a specific access list number to display this detail on a single IP access list. **SHOW ACCESS-LISTS** will not display what interface an access list has been applied to, as the **SHOW INTERFACE** commands do.

➤ *SHOW IP ACCESS-LIST*—This command will display all IP access lists in use on the router. It will also show each line of the access list and return the number of times a packet matched that line. This command can also be used with a specific IP access list number to display this detail on a single IP access list. **SHOW IP ACCESS-LISTS** will not display what interface an access list has been applied to, as the **SHOW INTERFACE** commands do.

➤ *SHOW IPX ACCESS-LIST*—This command will display all IPX access lists in use on the router. It will also show each line of the access list and return the number of times a packet matched that line. This command can also be used with a specific IPX access list number to display this detail on a single IPX access list. **SHOW IPX ACCESS-LISTS** will not display what interface an access list has been applied to, as the **SHOW INTERFACE** commands do.

➤ *SHOW IP INTERFACE [interface number]*—This command will display the IP access lists that have been applied to a specific interface. It will not display any matches to the individual lines of the access list, as the **SHOW ACCESS-LIST** commands do.

➤ *SHOW IPX INTERFACE [interface number]*—This command will display the IPX access lists that have been applied to a specific interface. Will not display any matches to the individual lines of the access list, as the **SHOW ACCESS-LIST** commands do.

For more information, see Chapter 13 of *CCNA Routing and Switching Exam Cram, Second Edition.*

Question 54

The correct answer is d. Two common protocols used when encapsulating data over serial links are SDLC and HDLC. SDLC is a data-link layer protocol used by IBM networks when communicating over WAN links using the Systems Network Architecture (SNA) protocol. HDLC, or high-level data-link control, is an ISO standard data-link protocol used in WAN communication. Cisco routers use HDLC as the default protocol for all serial (WAN) links. All of the other answers are invalid descriptions of the SDLC protocol.

For more information, see Chapter 4 of *CCNA Routing and Switching Exam Cram, Second Edition.*

Question 55

The correct answer is a. The OSI model defines standards used in networking and comprises a seven-layer model. The application layer of the OSI model functions as the "window" for applications to access network resources and is responsible for verifying the appropriate resources exist to make a connection. Since no other layer of the OSI model functions as the "window" for applications to access network resources, the other answers are incorrect. The following outlines the seven layers of the OSI model and their functions:

➤ *Application*—The "window" to networking used by programs, the application layer is responsible for:

  ➤ Verifying that the appropriate resources are present to initiate a connection with the destination node.

  ➤ Verifying the identity of the destination node.

Application layer protocols include Telnet, File Transfer Protocol (FTP), Simple Mail Transfer Protocol (SMTP), World Wide Web (WWW), Electronic Data Interchange (EDI) and Wide Area Information Server (WAIS).

➤ *Presentation*—Essentially a translator, the presentation layer is responsible for:

  ➤ Translating text and data syntax, such as Extended Binary-Coded Decimal Interchange Code (EBCDIC, used in IBM systems) to American Standard Code for Information Interchange (ASCII, used in PC and most computer systems).

➤ Using Abstract Syntax Notation to perform data translation.

Presentation layer protocols include PICTure (PICT, Apple computer picture format), Tagged Image File Format (TIFF), Joint Photographic Experts Group (JPEG), Musical Instrument Digital Interface (MIDI), Motion Picture Expert Group (MPEG) and Quick Time (audio/video application).

➤ *Session*—This is used to coordinate communication between nodes; the session layer is responsible for:

  ➤ Creating, maintaining, and ending a communication session.

  ➤ Coordinating service requests and responses that occur between nodes across the network.

Session layer protocols include Network File System (NFS, used by SUN Microsystems and Unix with TCP/IP), structured query language (SQL, used to define database information requests), Remote Procedure Calls (RPC, used in Microsoft network communication), X Window (used by Unix terminals), AppleTalk Session Protocol (ASP, used by Apple computers) and Digital Network Architecture Session Control Protocol (DNA SCP, used by IBM).

➤ *Transport*—Reliable end-to-end communication is the primary function of the transport layer. Its many responsibilities include:

  ➤ Ensuring flow control so the amount of data being sent from one node will not overwhelm the destination node.

  ➤ Ensuring the ability of multiple applications to utilize a single transport (multiplexing).

  ➤ Ensuring the reliable transfer of data. The transport layer implements connection-oriented services between nodes, which utilizes a three-way handshake (synchronization, acknowledgment, and data transfer) to efficiently transfer data.

  ➤ Ensuring positive acknowledgment, which is the process of a node waiting for an acknowledgment from the destination node prior to sending data.

  ➤ Windowing, which is a form of flow control that specifies how much data will be transferred between acknowledgments.

Transport layer protocols include Transmission Control Protocol (TCP, part of the TCP/IP protocol suite) and Sequenced Packet Exchange (SPX, part of the IPX/SPX protocol suite).

➤ *Network*—This selects the appropriate path a packet should take to get to its intended destination. It is responsible for routing, which is the process of using a network layer address to determine the best path a packet will travel to its destination. Network layer protocols include Internet Protocol (IP, part of the TCP/IP protocol suite), and Internetwork Packet Exchange (IPX, part of the IPX/SPX protocol suite).

➤ *Data Link*—Divided into two sublayers (Media Access Control [MAC], and Logical Link Control [LLC]), the data-link layer is responsible for:

> ➤ Preparing data from upper layers to be transmitted over the physical medium by encapsulating upper-layer data into frames. This frame includes the source and destination of MAC addresses in the frame header.

> ➤ Converting data into bits, so it can be transmitted by the physical layer.

> ➤ Adding a cyclical redundancy check (CRC) to the end of a frame, used for error checking at the data-link layer.

Data-link protocols include High-Level Data Link Control (HDLC, the Cisco default encapsulation for serial connections), Synchronous Data Link Control, (SDLC, used in IBM networks), Link Access Procedure Balanced (LAPB, used with X.25), X.25 (packet switching network), Serial Line Internet Protocol (SLIP, an older TCP/IP dial-up protocol), Point-to-Point Protocol (PPP, a newer dial-up protocol), Integrated Services Digital Network (ISDN, a dial-up digital service), and frame relay (a packet switching network).

➤ *Physical*—The lowest layer of the OSI model, the physical layer defines the electrical functionality required to send and receive bits over a given physical medium. Specifications that define the voltage levels and physical components of a network are defined at the physical layer. Protocols are not specified at the physical layer, because they are implemented as software. Examples of the standards for sending data over the physical medium are Ethernet (the most widely used standard in networking), Token Ring (IBM's proprietary networking topology), and Fiber Distributed Data Interface (FDDI, a standard for fiber optic networks, commonly used as a backbone).

For more information, see Chapter 4 of *CCNA Routing and Switching Exam Cram, Second Edition.*

Question 56

The correct answer is a. The access list, **ACCESS-LIST 800 DENY 15 25**, will deny IPX network 15 access to IPX network 25. Answer b is incorrect, because the source and destination networks are in the wrong order. The syntax for a standard IPX access list specifies the source first, then the destination. Answer c is incorrect, because the numerical range for a standard IPX access list is 800 through 899. Answer d is incorrect, because no hyphen is between **ACCESS** and **LIST**.

Access lists can be used to filter the traffic that is handled by a Cisco router. An IPX access list will filter IPX traffic and can be created as a standard IPX access list or an extended IPX access list. A standard IPX access list can be used to permit or deny access based on the source and destination IPX address information. The syntax for creating a standard IPX access list is:

```
ACCESS-LIST [access-list-number] [permit|deny] source
 [source-node-mask] [destination] [[destination-node-mask]
```

➤ *ACCESS-LIST*—This command is followed by the **access list number**. When using a standard IPX access list, this can be a number between 800 and 899.

➤ *[permit|deny]*—This statement will either allow or disallow traffic in this access list.

➤ *source*—This statement is used to specify the source IPX addressing information. A "-1" can be used to signify "all networks."

➤ *[destination]*—This statement is used to specify destination addressing information ("-1" may be used here as well).

➤ *[source-node-mask] and [[destination-node-mask]*—These are optional parameters for controlling access to a portion of a network, similar to the wildmask that is used in IP access lists.

For more information, see Chapter 13 of *CCNA Routing and Switching Exam Cram, Second Edition*.

Question 57

The correct answer is a. When configuring IP on a Cisco router interface, you must specify the IP address for the interface and the number of bits used in the subnet mask. When an IP address is specified for the interface, the router will identify the address by the first octet rule as either a Class A, B, or C address

(Class A=1–126, Class B=28–191, Class C=192–223). The default subnet mask for the address class is assumed to be present by the router (Class A=255.0.0.0, Class B=255.255.0.0, Class C=255.255.255.0).

The second step to configuring IP on a router interface involves specifying the number of bits in the subnet field. If this number is 0, the default subnet mask is applied based on the address class. The number of bits specified in the subnet field will be applied by the router to the host portion of the IP address. In this question, a Class B address was entered for the interface (130.200.200.254). The router then assumed a subnet mask of 255.255.0.0, the default subnet mask for a Class B address. The number of bits in the subnet field was specified as 7, which was interpreted by the router as 7 bits in the *third* octet of the IP address and is applied as 255.255.254.0. (Seven bits in binary = 11111110, which is the decimal equivalent to 254.) Answer b is incorrect, because the subnet mask is 255.255.254.0. Answer c is incorrect, because this is a Class B address, not a Class C address. Answer d is incorrect, because the subnet mask is 255.255.254.0.

For more information, see Chapter 8 of *CCNA Routing and Switching Exam Cram, Second Edition.*

Question 58

The correct answers are a and b. Both the **ENABLE** and **ENABLE SECRET** passwords can be used to control access to the privileged mode of a router. Answers c, d, and e are all valid types of passwords used in a Cisco router, but they do not control access to privileged mode. Answer f is not a valid password. Cisco routers utilize passwords to provide security access to a router. Passwords can be set for controlling access to privileged mode, access to remote sessions via Telnet, or access through the auxiliary port. The following list details the common passwords used in Cisco routers:

➤ *Enable Password*—The enable password is used to control access to the privileged mode of the router. If an enable password is specified, it must be entered after the **ENABLE** command to successfully access privileged mode. To configure the enable password, use the **ENABLE PASSWORD** [*password*] command.

➤ *Enable Secret Password*—The enable secret password is used to control access to privileged mode, similar to the enable password. The difference between enable secret and enable password is the enable secret password will be encrypted for additional security, and it will take precedence over the enable password if both are enabled. To configure the enable secret password, use the **ENABLE SECRET** [*password*] command.

➤ *Virtual Terminal Password*—The virtual terminal password is used to control remote Telnet access to a router. Setting the virtual terminal password will require all users that Telnet into the router to provide this password for access. To configure the virtual terminal password, use the following commands:

```
LINE VTY 0 4
LOGIN
PASSWORD [password]
```

➤ *Auxiliary Password*—The auxiliary password is used to control access to any auxiliary ports the router may have. Setting the auxiliary password will require all users connecting to the auxiliary port of the router (usually remote dial-in) to provide this password for access. To configure the auxiliary password, use the following commands:

```
LINE AUX 0
LOGIN
PASSWORD [password]
```

➤ *Console Password*—The console password is used to control access to the console port of the router. Setting the console password will require all users connecting to the router console to provide this password for access. To configure the console password, use the following commands:

```
LINE CON 0
LOGIN
PASSWORD [password]
```

For more information, see Chapter 8 of *CCNA Routing and Switching Exam Cram, Second Edition.*

Question 59

The correct answers are a and d. Frame relay is a packet switching technology that utilizes logical circuits to form a connection. This logical connection is identified by the DLCI. The DLCI is a unique identifier used to identify a specific frame relay connection. A frame relay connection is managed through the LMI signaling format, which provides information about the DLCI values, as well as other connection-related information. Your Cisco router must utilize the same LMI signaling format as your service provider. There are three LMI signaling formats: CISCO, ANSI, and Q933A. The Cisco default LMI type is CISCO. X.25 is a packet switching technology that also utilizes logical or virtual circuits to facilitate WAN connectivity. The two types of logical circuits used in X.25 are SVC and PVC. An SVC is a temporary connection that is established on an as-needed basis and is closed when not in use. A PVC is a permanent connection that is always available for use. Answers b and d are incorrect, because two types of logical circuits used in X.25 are SVC and PVC.

For more information, see Chapter 11 of *CCNA Routing and Switching Exam Cram, Second Edition.*

Question 60

The correct answers are a, c, and f. A bridge resides at the data-link layer of the OSI model and can filter traffic based on the data-link or MAC address of a packet. (The MAC address is the address burned into the network adapter card by the manufacturer.) Answer e is incorrect, because bridges pass unknown packets out all ports and also pass broadcast traffic. To prevent the potential loops that can occur from this broadcast traffic, bridges use the Spanning Tree Protocol, which allows for redundant connections between bridges while blocking traffic that can cause loops. Answer b is incorrect, because a router resides at the network layer of the OSI model and can filter traffic based on the network-layer address (IP address). Answer d is incorrect, because routers do not pass broadcast traffic and will discard any packets with unknown destinations.

For more information, see Chapter 4 of *CCNA Routing and Switching Exam Cram, Second Edition.*

Question 61

The correct answer is a. Flash is an erasable, programmable memory area that contains the Cisco operating system software (IOS). The internal components of Cisco routers are comparable to that of conventional PCs. Answer b is incorrect, because RAM does not store the Cisco IOS. Answer c is incorrect, because ROM is not erasable or programmable. If ROM needs to be modified, the chip must be replaced. Answer d is incorrect, because there is no such thing as Flash RAM. A Cisco router is similar to a PC in that it contains an operating system (known as the IOS), a bootstrap program (like the BIOS used in a PC), RAM (memory), and a processor (CPU). Unlike a PC, a Cisco router has no external drives to install software or upgrade the IOS. Rather than use a floppy or CD drive to enter software as PCs do, Cisco routers are updated by downloading update information from a TFTP server. The following list outlines the core components of Cisco routers:

➤ *Read-Only Memory (ROM)*—This is a physical chip installed on a router's motherboard that contains the bootstrap program, the POST, and the operating system software (Cisco IOS). When a Cisco router is first powered up, the bootstrap program and the POST are executed. ROM cannot be changed through software; the chip must be replaced if any modifications are needed.

➤ *Random Access Memory (RAM)*—As with PCs, RAM is used as the router's main working area and contains the running configuration. All information in RAM dissipates when the router is turned off.

➤ *Non-Volatile RAM (NVRAM)*—This is used by Cisco routers to store the startup or running configuration. After a configuration is created, it exists and runs in RAM. Because RAM cannot permanently store this information, NVRAM is used. Information in NVRAM is preserved after the router is powered off.

➤ *Flash*—Essentially, this is the PROM used in PCs. The Flash is an erasable, programmable memory area that contains the Cisco operating system software (IOS). When a router's operating system needs to be upgraded, new software can be downloaded from a TFTP server to the router's Flash. Upgrading the IOS in this manner is typically more convenient than replacing the ROM chip in the router's motherboard. Information in Flash is retained when the router is powered off.

For more information, see Chapter 4 of *CCNA Routing and Switching Exam Cram, Second Edition.*

Question 62

The correct answer is a. User mode is used to display basic router system information and connect to remote devices. The commands in user mode are limited in capability, and when you first log into a router, you are placed in user mode. The router prompt is followed by an angle bracket (**ROUTER>**). All of the other answers are incorrect, because they are all modes that offer extended capabilities. The Cisco IOS uses a command interpreter known as EXEC, which contains two primary modes: user mode and privileged mode. Privileged mode is used to modify and view the configuration of the router, as well as to set IOS parameters. This mode also contains the **CONFIGURE** command, which is used to access other configuration modes, such as global configuration mode and interface mode. In privileged mode, the router prompt is followed by a pound sign (**ROUTER#**). To enter privileged mode, use the **ENABLE** command from user mode.

For more information, see Chapter 8 of *CCNA Routing and Switching Exam Cram, Second Edition.*

Question 63

The correct answer is b. The **CONFIGURE MEMORY** command can be abbreviated as **CONFIG MEM** and must be entered from privileged mode where the router prompt is followed by a pound sign (**ROUTER#**). Answer a is incorrect, because the router is in user mode, and the **CONFIGURE MEM** command cannot be entered from user mode. In this mode, the router prompt is followed by an angle bracket (**ROUTER>**). Answer c is incorrect, because the prompt is already in configuration mode. After entering configuration mode, the router prompt displays "config" in parentheses (**ROUTER(config)#**). Answer d is incorrect, because the prompt is displaying interface mode (**ROUTER(CONFIG-IF)#**).

For more information, see Chapter 8 of *CCNA Routing and Switching Exam Cram, Second Edition.*

Question 64

The correct answer is c. Access lists can be used to filter the traffic that is handled by a Cisco router. An IP access list will filter IP traffic and can be created as a standard IP access list or an extended IP access list. A standard IP access list can be used to permit or deny access based on IP addressing information only and can only act on the source IP addressing information. Since a standard IP access list can only act on the source IP address, all of the other answers are incorrect. An extended IP access list can be used to permit or deny access based on the source and destination IP address, the protocol, or the port number.

For more information, see Chapter 13 of *CCNA Routing and Switching Exam Cram, Second Edition.*

Question 65

The correct answer is a. The Tab key will complete an entry typed at the router prompt. For example, typing "**CONFIG**" and pressing Tab will result in **CONFIGURE** being displayed. The other answers are not valid uses for the Tab key. Configuring a Cisco router can sometimes involve long, detailed command lines that can be slow to navigate. For this reason, the Cisco IOS includes a number of shortcuts through them. The following list details some of the more common methods used to navigate the command line:

➤ *Ctrl+A*—Pressing the Control key along with the letter *A* will move the cursor to the beginning of the command line.

➤ *Ctrl+E*—Pressing the Control key along with the letter *E* will move the cursor to the end of the command line.

➤ *Ctrl+P*—Pressing the Control key along with the letter *P* will recall the previous command that was entered. (The up arrow will do this as well.)

➤ *Ctrl+N*—Pressing the Control key along with the letter *N* will move to the most recent command that was entered. (The down arrow will do this as well.)

➤ *Esc+B*—Pressing the Escape key along with the letter *B* will move the cursor back one word.

➤ *Esc+F*—Pressing the Escape key along with the letter *F* will move the cursor forward one word.

➤ *Left and Right Arrow Keys*—These arrow keys will move the cursor one character left and right, respectively.

For more information, see Chapter 8 of *CCNA Routing and Switching Exam Cram, Second Edition.*

Question 66

The correct answer is b. Pressing the Control key along with the letter *E* will move the cursor to the end of the command line. Since this is the only use for the Ctrl+E key, all of the other answers are incorrect. Configuring a Cisco router can sometimes involve long, detailed command lines that can be slow to navigate. For this reason, the Cisco IOS includes a number of shortcuts through them. The following list details some of the more common methods used to navigate the command line:

➤ *Ctrl+A*—Pressing the Control key along with the letter *A* will move the cursor to the beginning of the command line.

➤ *Ctrl+E*—Pressing the Control key along with the letter *E* will move the cursor to the end of the command line.

➤ *Ctrl+P*—Pressing the Control key along with the letter *P* will recall previous commands in the command history. (The up arrow will do this as well.)

➤ *Ctrl+N*—Pressing the Control key along with the letter *N* will return to more recent commands in the command history after recalling previous commands. (The down arrow will do this as well.)

➤ *Esc+B*—Pressing the Escape key along with the letter *B* will move the cursor back one word.

➤ *Esc+F*—Pressing the Escape key along with the letter *F* will move the cursor forward one word.

➤ *Left and Right Arrow Keys*—These arrow keys will move the cursor one character left and right, respectively.

➤ *Tab*—Pressing the Tab key will complete an entry typed at the router prompt.

For more information, see Chapter 8 of *CCNA Routing and Switching Exam Cram, Second Edition.*

Question 67

The correct answer is c. **COPY STARTUP-CONFIG RUNNING-CONFIG** will copy the startup configuration file in NVRAM to the running configuration in RAM. Answer a is incorrect, because **COPY STARTUP-CONFIG TO RUNNING-CONFIG** is not a valid Cisco IOS command. Answer b is incorrect, because the correct syntax of the **COPY STARTUP-CONFIG RUNNING-CONFIG** contains a hyphen between **RUNNING** and **CONFIG**. Answer d is incorrect, because **COPY RUNNING-CONFIG FROM STARTUP CONFIG** is not a valid Cisco IOS command. The startup configuration is a configuration file in a Cisco router that contains the commands used to set router-specific parameters. The startup configuration is stored in NVRAM and is loaded into RAM when the router starts up.

The running configuration is the startup configuration that is loaded into RAM, and is the configuration used by the router when it is running. Because any information in RAM is lost when the router is powered off, the running configuration can be saved to the startup configuration, which is located in NVRAM and will, therefore, be saved when the router is powered off. In addition, both the running and startup configuration files can be copied to and from a TFTP server, for backup purposes. The following list outlines the common commands used when working with configuration files:

➤ *COPY STARTUP-CONFIG RUNNING-CONFIG*—This command will copy the startup configuration file in NVRAM to the running configuration in RAM.

➤ *COPY STARTUP-CONFIG TFTP*—This command will copy the startup configuration file in NVRAM to a remote TFTP server.

➤ *COPY RUNNING-CONFIG STARTUP-CONFIG*—This command will copy the running configuration from RAM to the startup configuration file in NVRAM.

➤ *COPY RUNNING-CONFIG TFTP*—This command will copy the running configuration from RAM to a remote TFTP server.

➤ *COPY TFTP RUNNING-CONFIG*—This command will copy a configuration file from a TFTP server to the router's running configuration in RAM.

➤ *COPY TFTP STARTUP-CONFIG*—This command will copy a configuration file from a TFTP server to the startup configuration in NVRAM.

Note that the basic syntax of the **COPY** command specifies the source first, then the destination.

To modify a router's running configuration, you must be in global configuration mode. The **CONFIG T** command is used in privileged mode to access global configuration mode. Global configuration mode is displayed at the router prompt with "config" in parentheses **(RouterA(config)#)**. To exit global configuration mode, use the keys Control and Z (Ctrl+Z).

For more information, see Chapter 8 of *CCNA Routing and Switching Exam Cram, Second Edition.*

Question 68

The correct answer is d. The display shows the router is at the user mode prompt **(ROUTER>)**, and the **ENABLE** command has just been issued. The "Password" prompt signifies that an enable or enable secret password has been set for access to privileged mode. Both the enable and the enable secret passwords are used to control access to privileged mode. The enable secret password will take precedence over the enable password if both are enabled. The enable password is used to control access to the privileged mode of the router. If an enable password is specified, it must be entered after the **ENABLE** command to successfully access privileged mode.

Answer a is incorrect, because the console password is used to control access to the console port of the router. Answer b is incorrect, because the virtual terminal password is used to control remote Telnet access to a router. Setting the virtual terminal password will require all users that Telnet into the router to provide this password for access. Answer c is incorrect, because the auxiliary password is used to control access to any auxiliary ports the router may have.

Setting the auxiliary password will require all users connecting to the auxiliary port of the router (usually remote dial-in) to provide this password for access. The enable password is used to control access to the privileged mode of the router. If an enable password is specified, it must be entered after the **ENABLE** command to successfully access privileged mode.

For more information, see Chapter 8 of *CCNA Routing and Switching Exam Cram, Second Edition.*

Question 69

The correct answer is c. Access lists can be used to filter the traffic that is handled by a Cisco router. An IP access list will filter IP traffic and can be created as a standard IP access list or an extended IP access list. A standard IP access list can be used to permit or deny access based on IP addressing information only and can only act on the source IP addressing information. An extended IP access list can be used to permit or deny access based on the source and destination IP address, the protocol, or the port number. Answers a and b are incorrect, because an extended IP access list can act on more than just the source IP address. Answer d is incorrect, because a standard IP access list can only act on the source IP address.

For more information, see Chapter 13 of *CCNA Routing and Switching Exam Cram, Second Edition.*

Question 70

The correct answers are a, b, and d. RIP is a distance vector routing protocol used to automatically update entries in a routing table. Answer c is incorrect, because RIP only uses hop count to select the best path. Answer e is incorrect, because the maximum hop count used in RIP is 15. Any destination that requires more than 15 hops is considered unreachable by RIP. Answer f is incorrect, because RIP routers update their routing tables by broadcasting the contents of their routing table every 30 seconds. This broadcast traffic does not make RIP very efficient for large networks. Answer g is incorrect, because RIP is a distance vector routing protocol, not a link state routing protocol.

For more information, see Chapter 11 of *CCNA Routing and Switching Exam Cram, Second Edition.*

Practice Test #3

Question 1

Consider the following access list:

```
ACCESS-LIST 110 PERMIT TCP HOST 77.90.50.22
HOST 210.34.5.20 EQ 25
```

Based on this access list, which of the following statements is true?

○ a. This access list will allow SMTP traffic from host 210.34.5.20 to host 77.90.50.22.

○ b. This access list will allow FTP traffic from host 77.90.50.22 to host 210.34.5.20.

○ c. This access list will allow SMTP traffic from host 77.90.50.22 to host 210.34.5.20.

○ d. This access list will allow SMTP traffic from host 210.34.5.20 to host 77.90.50.22.

Question 2

Which of the following protocols reside at the network layer of the OSI model? [Choose the two best answers]

❑ a. IPX

❑ b. SPX

❑ c. TCP

❑ d. IP

❑ e. PICT

❑ f. RPC

❑ g. ASP

Question 3

Your Cisco router console displays the following:

```
Router con0 is now available
Press RETURN to get started.
Router>enable
Password:
Router#CONFIG T
Enter configuration commands, one per line.
 End with CRTL/Z.
Router(config)#
```

Based on this display, what mode is this router currently in?

○ a. The router is in user mode.

○ b. The router is in privileged mode.

○ c. The router is in global configuration mode.

○ d. There is not enough information to answer the question.

Question 4

Seated at the router console, you wish to change the setting for the clock. You know the correct command to use is "**CLOCK**," but you are not sure of the keywords that are used with this command. Which of the following best describes the correct method you could use to find out what keywords can be used with the **CLOCK** command?

- ○ a. Type "CLOCK?" at the router prompt.
- ○ b. Type "CLOCK ?" at the router prompt.
- ○ c. Type "HELP CLOCK" at the router prompt.
- ○ d. Type "SHOW CLOCK*" at the router prompt.

Question 5

The Cisco IOS includes a command history feature that can be used to store previously entered commands in a history buffer. Which of the following commands would you enter to display all of the commands currently stored in the history buffer?

- ○ a. **SHOW HISTORY**
- ○ b. **SHOW TERMINAL HISTORY**
- ○ c. **DISPLAY HISTORY**
- ○ d. **HISTORY**

Question 6

Your co-worker, Annie, is having trouble displaying the startup configuration file on her Cisco router and asks for your help. You look at the router console and observe the following display:

```
Router con0 is now available
Press RETURN to get started.
Router>SHOW STARTUP-CONFIG
                ^
% Invalid input detected at '^' marker.
Router>
```

Based on this display, what is the most likely reason Annie cannot display the router's startup configuration?

○ a. Annie entered the wrong command. The correct command is **SHOW STARTUP**.

○ b. Annie is in the privileged mode of the router. The router needs to be in user mode for this command to work.

○ c. Annie is in user mode of the router. The router needs to be in privileged mode for this to work.

○ d. Annie entered the wrong command. The correct command is **SHOW STARTUP CONFIG.**

Question 7

You have configured several IP access lists on your Cisco router. You need to see what matches have occurred for IP access list number 127. Which of the following commands can you use to see what lines in IP access list 127 were matched by the traffic that passed through the router?

○ a. **SHOW IP ACCESS LIST 127**

○ b. **SHOW IP INTERFACE 127**

○ c. **SHOW IP ACCESS-LIST 127**

○ d. **SHOW IP INTERFACE 127**

Question 8

Consider the following IPX address:

```
2c.1234.1111.2222
```

Based on this information, what is the network ID of this address?

○ a. 2c.1234

○ b. 1234.1111.2222

○ c. 1111.2222

○ d. 2c

Question 9

You are troubleshooting a Cisco router. The router is using TCP/IP as the only protocol. You need to see the routing protocols in use for IP on the entire router, as well as update frequency and filter information. Which of the following Cisco IOS commands can you use to view this information?

○ a. **SHOW IP PROTOCOL**

○ b. **SHOW IP ROUTE**

○ c. **SHOW IP-PROTOCOL**

○ d. **SHOW IP-ROUTE**

Question 10

Two nodes on a network are communicating. One is using the American Standard Code for Information Interchange (ASCII) format, and the other is using the Extended Binary-Coded Decimal Interchange Code (EBCDIC) format. Which layer of the OSI model would perform the data syntax translation between ASCII and EBCDIC?

○ a. Application

○ b. Presentation

○ c. Session

○ d. Transport

○ e. Network

○ f. Data link

○ g. Physical

Question 11

Which of the following statements best describes the use of the **CONFIGURE OVERWRITE-NETWORK (CONFIG O)** command?

- ○ a. **CONFIGURE OVERWRITE-NETWORK** is used to enter configuration commands into the router from the console port or through Telnet.

- ○ b. **CONFIGURE OVERWRITE-NETWORK** is used to copy the configuration file from a TFTP server into the router's RAM.

- ○ c. **CONFIGURE OVERWRITE-NETWORK** is used to copy a configuration file into NVRAM from a TFTP server.

- ○ d. **CONFIGURE OVERWRITE-NETWORK** is used to execute the configuration stored in NVRAM and will copy the startup configuration to the running configuration.

Question 12

Which of the following correctly states the use of Ctrl+P at the command line in a Cisco router?

- ○ a. Ctrl+P will move the cursor to the beginning of the command line.

- ○ b. Ctrl+P will recall previous commands in the command history.

- ○ c. Ctrl+P will move the cursor back one word.

- ○ d. Ctrl+P will return to more recent commands in the command history after recalling previous commands.

Question 13

In an effort to provide security on your Cisco router, you want to require a password to access the privileged mode of the router. You want the password to be "PENNICK". Additionally, you want the password to be encrypted to prevent unauthorized users from capturing it. Which of the following is the correct Cisco IOS command you would issue that will require the password "PENNICK" to enter privileged mode and will also encrypt this password when it is entered?

- ○ a. **ENABLE PASSWORD PENNICK**

- ○ b. **ENABLE SECRET PASSWORD PENNICK**

- ○ c. **LINE ENABLE SECRET PENNICK**

- ○ d. **ENABLE SECRET PENNICK**

Question 14

You have been assigned a Class C IP address for use in your network. You will use subnet mask 255.255.255.192.

Based on this information, how many total bits are available for host IDs on this network?

○ a. 6

○ b. 2

○ c. 24

○ d. 26

Question 15

Encapsulation involves five steps. First, user information is converted to data. Second, data is converted to segments. Third, segments are converted to packets or datagrams. What is the fourth step of encapsulation?

○ a. Packets and datagrams are converted to frames.

○ b. Packets and datagrams are converted to bits.

○ c. Packets and datagrams are converted to segments.

○ d. Segments are converted to frames.

Question 16

You have configured several IPX access lists on your Cisco router. Which of the following commands can you use to see a list of all the IPX access lists that are currently in use in your router?

○ a. **SHOW IPX ACCESS LIST**

○ b. **SHOW IPX ACCESS-LIST**

○ c. **SHOW-IPX-ACCESS LIST**

○ d. **SHOW IPX ACCESS LIST 800-899**

Question 17

Seated at your router console, you create the following access list:

```
ACCESS LIST 10 PERMIT 200.200.200.75
```

You want to apply this access list to the outgoing serial interface number 1 of your router. You enter the interface configuration mode of your router and access serial interface 1.

Which of the following commands will you enter to perform this operation?

○ a. **IP ACCESS-GROUP 10 OUT**

○ b. **IP ACCESS GROUP 10 OUT**

○ c. **ACCESS-GROUP 10 OUT**

○ d. **IP ACCESS-GROUP SO OUT**

Question 18

You need to configure a static route on your Cisco router. You want the router to send all packets destined for network 131.200.0.0 255.255.0.0 to router 220.200.200.84. Which of the following is the correct command to create a static route that will perform this operation?

○ a. **IP-ROUTE 131.200.0.0 255.255.0.0 220.200.200.84**

○ b. **STATIC ROUTE 131.200.0.0 255.255.0.0 220.200.200.84**

○ c. **IP ROUTE 131.200.0.0 255.255.0.0 220.200.200.84**

○ d. **IP ROUTE 220.200.200.84 255.255.0.0 131.200.0.0**

Question 19

A problem that can occur with distance vector routing protocols are routing loops. One of the methods used by distance vector routing protocols to prevent routing loops is maximum hop count. Which of the following statements correctly describes the function of Maximum Hop Count?

○ a. Maximum hop count involves setting a maximum hop count that a packet can travel before it is discarded.

○ b. Maximum hop count is when a router poisons a route by setting a hop count of 16.

○ c. Maximum hop count states that a packet cannot be sent back in the direction that it was received.

○ d. Maximum hop count is a specified amount of time a router will wait before updating its routing table.

Question 20

You observe the following configuration on one of the computers in your TCP/IP network:

```
IP address : 150.150.33.90
Subnet Mask 255.255.0.0
Default Gateway : 150.150.11.5
```

Based on this information, which of the following statements are true? [Choose the three best answers]

❏ a. The IP address is a Class B address.

❏ b. The IP address is a Class A address.

❏ c. The Network ID of IP address 150.150.33.90 is 150.150.33.0.

❏ d. The Network ID of IP address 150.150.33.90 is 150.150.0.0.

❏ e. The Host ID of IP address 150.150.33.90 is 150.33.0.

❏ f. The Host ID of IP address 150.150.33.90 is 33.90.

Question 21

Which of the following protocols does not reside at the session layer of the OSI model?

○ a. RPC

○ b. NFS

○ c. DNA SCP

○ d. SQL

○ e. X Window

○ f. ASP

○ g. HDLC

Question 22

Which layer of the OSI model adds a CRC, or Cyclical Redundancy Check, to the end of a frame that is used for error checking?

○ a. Application

○ b. Presentation

○ c. Session

○ d. Transport

○ e. Network

○ f. Data-link

○ g. Physical

Question 23

Which of the following is the correct numerical range used for a standard IPX access list?

○ a. 1–99

○ b. 100–199

○ c. 800–899

○ d. 900–999

Question 24

Which layer of the OSI model is responsible for preparing data from upper layers to be transmitted over the physical medium by encapsulating data into frames?

- ○ a. Application
- ○ b. Presentation
- ○ c. Session
- ○ d. Transport
- ○ e. Network
- ○ f. Data-link
- ○ g. Physical

Question 25

Which of the following best describes the function and contents of the Flash in a Cisco router?

- ○ a. Flash is a physical chip installed on a router's motherboard that contains the bootstrap program, the POST (Power On Self Test), and the operating system software (Cisco IOS).
- ○ b. Flash is used by Cisco routers to store the startup configuration.
- ○ c. Flash is an erasable, programmable memory area that contains the Cisco operating system software (IOS).
- ○ d. Flash is used as the router's main working area and contains the running configuration.

Question 26

Frame relay is a packet switching WAN technology that utilizes logical or virtual circuits to form a connection. Each frame relay virtual circuit is assigned a unique identifier used when transmitting packets across the WAN. Which of the following is the correct term used to describe how a logical circuit in a frame relay network is identified?

- ○ a. The logical circuit in a frame relay network is identified by the American National Standards Institution (ANSI).

- ○ b. The logical circuit in a frame relay network is identified by the data-link connection identifier (DLCI).

- ○ c. The logical circuit in a frame relay network is identified by the switched virtual circuits (SVC).

- ○ d. The logical circuit in a frame relay network is identified by the permanent virtual circuits (PVC).

Question 27

Cisco routers support the use of access lists, which can be used to filter incoming or outgoing traffic. Each type of access list is assigned a range of numbers that distinguish one access list type from another. Which of the following are the correct numerical ranges for access lists? [Choose the four best answers]

- ❑ a. A standard IP access list uses numbers between 1 and 99.

- ❑ b. A standard IP access list uses numbers between 0 and 99.

- ❑ c. An extended IP access list uses numbers between 1 and 99.

- ❑ d. An extended IP access list uses numbers between 100 and 199.

- ❑ e. A standard IPX access list uses numbers between 800 and 899.

- ❑ f. A standard IPX access list uses numbers between 900 and 999.

- ❑ g. An extended IPX access list uses numbers between 900 and 999.

Question 28

A Cisco router's startup sequence is divided into three basic operations. First, the bootstrap program in ROM executes the POST (Power On Self Test), which performs a basic hardware level check. Second, the Cisco IOS (operating systems software) is loaded into memory per instructions from the boot system command. What is the next step in this startup sequence?

- ○ a.   The router's configuration file is loaded into memory from NVRAM.

- ○ b.   A second, more exhaustive, POST (Power On Self Test) executes.

- ○ c.   The router's configuration file is loaded into memory from ROM.

- ○ d.   The Flash performs a basic software level check.

Question 29

Which layer of the OSI model is responsible for verifying the identity of the destination node?

- ○ a.   Application
- ○ b.   Presentation
- ○ c.   Session
- ○ d.   Transport
- ○ e.   Network
- ○ f.   Data-link
- ○ g.   Physical

Question 30

Both a bridge and a router can be used to segment a network and improve network performance. Which of the following statements are true about the difference between a bridge and a router? [Choose the three best answers]

❑ a. A bridge resides at the network layer of the OSI model and can filter traffic based on the network layer address (IP address).

❑ b. A router resides at the network layer of the OSI model and can filter traffic based on the network layer address (IP address).

❑ c. Bridges pass unknown packets out all ports and also pass broadcast traffic.

❑ d. To prevent potential loops that can occur from this broadcast traffic, bridges use the Spanning Tree Protocol.

❑ e. Bridges do not pass broadcast traffic and will discard any packets with unknown destinations.

Question 31

You have just modified the running configuration of your Cisco router. Shortly after you did this, there was a brief power outage. Upon restarting the router, you notice that the changes you made to the configuration have been lost. Which of the following statements best describes why this occurred, and what you could have done to prevent it from happening?

○ a. The running configuration is stored in RAM and is not saved when the router is turned off. You could have saved the configuration file by using the **COPY STARTUP-CONFIG RUNNING-CONFIG** command.

○ b. The running configuration is stored in NVRAM and is not saved when the router is turned off. You could have saved the configuration file by using the **COPY RUNNING-CONFIG STARTUP-CONFIG** command.

○ c. The running configuration is stored in Flash and is not saved when the router is turned off. You could have saved the configuration file by using the **COPY RUNNING-CONFIG STARTUP CONFIG** command.

○ d. The running configuration is stored in RAM and is not saved when the router is turned off. You could have saved the configuration file by using the **COPY RUNNING-CONFIG STARTUP-CONFIG** command.

Question 32

A dynamic routing table is one that has its entries entered automatically by the router, through the use of a routing protocol. A dynamic routing protocol provides for automatic discovery of routes and eliminates the need for static routes. Routing protocols generally fall into one of two categories: link state and distance vector. Which of the following are considered distance vector routing protocols? [Choose the two best answers]

❏ a. RIP

❏ b. OSPF

❏ c. EIGRP

❏ d. IGRP

❏ e. IP

❏ f. IPX

Question 33

Which of the following protocols does not reside at the presentation layer of the OSI model?

○ a. JPEG

○ b. MPEG

○ c. Quick Time

○ d. WAIS

○ e. PICT

○ f. MIDI

○ g. TIFF

Question 34

Which of the following access lists will deny IPX network 33 access to IPX network 22?

○ a. **IPX ACCESS-LIST 888 DENY 33 22**

○ b. **ACCESS-LIST 810 DENY 22 33**

○ c. **ACCESS-LIST 100 DENY 33 22**

○ d. **ACCESS-LIST 888 DENY 33 22**

Question 35

Seated at your router console, you create the following extended IPX access list:

```
ACCESS LIST 922 DENY -1 22 0 37 0
ACCESS LIST 922 PERMIT -1 -1 0 -1 0
```

You want to apply this access list to the outgoing serial interface number 0 of your router. You enter the interface configuration mode of your router and access serial interface 0.

Which of the following commands will you enter to perform this operation?

○ a. **IPX ACCESS GROUP 922 OUT**

○ b. **IPX ACCESS-GROUP 922 OUT**

○ c. **ACCESS-GROUP 922 OUT**

○ d. **IPX ACCESS-LIST 922 OUT**

Question 36

Which of the following is a feature of connectionless protocols?

○ a. Creation of a virtual circuit.

○ b. Overhead is less than with connection-oriented protocols.

○ c. A communication session is established between hosts before sending data.

○ d. Sequencing.

Question 37

Which of the following is not a method used in flow control?

○ a. Multiplexing

○ b. Buffering

○ c. Source-quench messages

○ d. Windowing

Question 38

Although there are many different types of Cisco routers, they all share the same basic functionality. Which of the following are the four core internal components of a Cisco router? [Choose the four best answers]

❏ a. RAM

❏ b. Floppy drive

❏ c. NVRAM

❏ d. CD-ROM drive

❏ e. Flash

❏ f. ROM

❏ g. Ultra-SCSI drive

Question 39

Which of the following exec modes are you in when you first log into a router?

○ a. User mode

○ b. Privileged mode

○ c. Executive mode

○ d. Global configuration mode

Question 40

Which of the following is the correct command and prompt used to enter the EXEC privileged mode of a Cisco router?

○ a. **ROUTER>ENABLE**

○ b. **ROUTER>CONFIGURE**

○ c. **ROUTER#ENABLE**

○ d. **ROUTER#CONFIGURE**

Question 41

Which of the following is the correct command and prompt used to enter the **CONFIGURE NETWORK** command?

○ a. **ROUTER> CONFIG NET**

○ b. **ROUTER#CONFIG NET**

○ c. **ROUTER(config)#CONFIG NET**

○ d. **ROUTER(CONFIG-IF)#CONFIG NET**

Question 42

Consider the following display on your Cisco router:

```
Router con0 is now available
Press RETURN to get started.
Hello CCNA candidates!
Router>
```

Based on this display, which of the following commands was used to create the message "Hello CCNA candidates!"?

○ a. **CREATE MESSAGE**

○ b. **MAKE BANNER**

○ c. **BANNER MOTD**

○ d. **SET MESSAGE**

Question 43

You are monitoring the traffic filtered by IPX access list number 800 on your Cisco router. After making some changes to the internetwork, you decide to recheck the traffic matches to this access list again. Before you take these new readings, you would like to clear the previous matches from access list 800. Which of the following Cisco IOS commands could you use to perform this operation?

○ a. **CLEAR IPX ACCESS-LIST-COUNTERS 800**

○ b. **CLEAR ACCESS-LIST MATCHES 800**

○ c. **CLEAR ACCESS-LIST COUNTERS 800**

○ d. **CLEAR ACCESS LIST COUNTERS 800**

Question 44

When wide area networks communicate over telecommunication lines, several components are used to facilitate the connection. Which of the following best describes the CO, as it is used in WAN communication?

○ a. The CO, or company office, is the collection of communication devices (telephones, modems, and so forth) that exists at the customer's location.

○ b. The CO, or customer office, is the cabling that extends from the DEMARC to the telephone company's office.

○ c. The CO, or central office, is the telephone company's office that acts as the central communication point for the customer.

○ d. The CO, or calling office, is the point at the customer's premises where CPE devices connect and is usually a large punch-down board located in a wiring closet.

Question 45

A switch resides at the data-link layer of the OSI model and can filter traffic based on the data-link address of a packet or MAC address. (The MAC address is the address burned into the network adapter card by the manufacturer.) Cisco switches employ two basic methods of forwarding packets: Cut-Through and store-and-forward. Which switching method will copy an incoming packet to the local buffer and perform error checking on the packet before sending it on to its destination?

○ a. Cut-forward switching

○ b. Store-and-cut switching

○ c. Cut-through switching

○ d. Store-and-forward switching

Question 46

Which layer of the OSI model defines the electrical functionality required to send and receive bits over a given physical medium?

○ a. Application

○ b. Presentation

○ c. Session

○ d. Transport

○ e. Network

○ f. Data-link

○ g. Physical

Question 47

Which of the following statements are true about data-link and network layer addresses? [Choose the three best answers]

❑ a. Data-link or MAC addresses are "burned" into every network adapter card by the manufacturer.

❑ b. An example of a network layer address is an IP address.

❑ c. An example of a network layer address is a MAC address.

❑ d. An example of a data-link layer address is a MAC address.

Question 48

Consider the following display on your router console:

```
Router#CONFIG T
Enter configuration commands, one per line.
 End with CRTL/Z.
Router(config)#LINE VTY 04
Router(config-line)#LOGIN
Router(config-line)#
```

Assuming you want to set the virtual terminal password to be "EXAM", what would be the next command you would enter at the router prompt?

O a. **PASSWORD EXAM**

O b. **SET PASSWORD EXAM**

O c. **EXAM**

O d. **SET LOGIN PASSWORD EXAM**

Question 49

You are troubleshooting a Cisco router and need to make some specific changes to serial port 1. Which of the following Cisco IOS commands could you issue to access the configuration mode of serial port 1?

O a. **Router#interface serial 0**

O b. **Router(config)#interface serial 0**

O c. **Router>interface serial 0**

O d. **Router#configure interface serial 0**

Question 50

Which of the following is a valid standard IP access list?

O a. **ACCESS-LIST 100 PERMIT 131.200.0.0**

O b. **ACCESS-LIST 10 PERMIT 131.200.0.0 0.0.255.255255.255.0.0**

O c. **ACCESS LIST 10 PERMIT 131.200.0.0**

O d. **ACCESS-LIST PERMIT 131.200.0.0 255.255.0.0**

Question 51

Which of the following access lists will allow host 200.200.200.50 to have Telnet access to network 131.200.0.0?

- ○ a. **ACCESS-LIST 100 PERMIT TCP 200.200.200.50 0.0.0.0 131.200.0.0 0.0 255.255 EQ 23**

- ○ b. **ACCESS-LIST 10 PERMIT TCP 200.200.200.50 0.0.0.0 131.200.0.0 0.0.255.255 EQ 23**

- ○ c. **ACCESS-LIST 100 PERMIT TCP 200.200.200.50 0.0.0.0 131.200.0.0 0.0.255.255 EQ 21**

- ○ d. **ACCESS LIST 100 PERMIT TCP 200.200.200.50 0.0.0.0 131.200.0.0 0.0.255.255 EQ 23**

Question 52

The IPX/SPX protocol is used in Novell NetWare networks. When IPX packets are passed to the data-link layer, they are encapsulated into frames for transmission over the physical media. Novell NetWare supports a number of different encapsulation methods, or frame types, used to encapsulate IPX packets. The frame type used by NetWare computers must be the same as the one used by a Cisco router, or they will not be able to communicate. When configuring a Cisco router to support IPX, you can specify the frame type to use. Novell NetWare and Cisco both use different names to describe the same frame type. Which of the following statements are true about the frame types used by Novell and Cisco? [Choose the three best answers]

- ❑ a. The ETHERNET II frame type used by Novell is called SAP in Cisco.

- ❑ b. The ETHERNET 802.2 frame type used by Novell is called SAP in Cisco.

- ❑ c. The ETHERNET_snap frame type used by Novell is called SNAP in Cisco.

- ❑ d. The ETHERNET 802.3 frame type used by Novell is called NOVELL-ETHER in Cisco and is the default frame type.

- ❑ e. The TOKEN RING frame type used by Novell is called SNAP in Cisco.

- ❑ f. The TOKEN RING_SNAP frame type used by Novell is called ARPA in Cisco.

Question 53

The Cisco IOS includes a command history feature that can be used to store previously entered commands in a history buffer. Which of the following statements are true about the command history feature in Cisco routers? [Choose the two best answers]

❑ a. The command history buffer can be configured to hold up to 256 commands.

❑ b. The command history feature is enabled by default on Cisco routers.

❑ c. The command history feature is disabled by default on Cisco routers.

❑ d. The command history buffer holds 256 entries by default.

Question 54

You have just made changes to your Cisco router's running configuration and saved it to the startup configuration. You now want to have a backup copy of the startup configuration stored on a remote TFTP server. Which of the following is the correct Cisco IOS command to perform this operation?

○ a. **COPY STARTUPCONFIG TFTP**

○ b. **COPY STARTUP-CONFIG TFTP**

○ c. **COPY TFTP STARTUPCONFIG**

○ d. **COPY TFTP STARTUP-CONFIG**

Question 55

You are configuring IP on serial interface **s1** of your Cisco router. You specify the following at the router console:

```
Configuring interface Serial1:
Is this interface in use? [no]: y
Configure IP on this interface? [no]: y
Configure IP unnumbered on this interface?
 [no]: n
IP address for this interface: 130.200.200.254
Number of bits in subnet field [0]: 11
```

Based on this information, which of the following statements is true?

○ a. This is a Class B address, and the subnet mask is 255.255.224.0.

○ b. This is a Class B address, and the subnet mask is 255.255.255.224.

○ c. This is a Class C address, and the subnet mask is 255.255.255.224.

○ d. This is a Class B address, and the subnet mask is 255.255.0.0.

Question 56

Which of the following Cisco IOS commands are used to copy a configuration file into NVRAM from a TFTP server without altering the running configuration?

○ a. **CONFIGURE TERMINAL (CONFIG T)**

○ b. **CONFIGURE OVERWRITE-NETWORK (CONFIG O)**

○ c. **CONFIGURE NETWORK (CONFIG NET)**

○ d. **CONFIGURE MEMORY (CONFIG MEM)**

Question 57

Which of the following protocols reside at the transport layer of the OSI model? [Choose the two best answers]

❏ a. SMTP

❏ b. SPX

❏ c. Telnet

❏ d. SQL

❏ e. PICT

❏ f. RPC

❏ g. TCP

Question 58

Seated at the console of your Cisco router, you wish to display the current running configuration. Which of the following Cisco IOS commands could you enter to perform this operation? [Choose the two best answers]

❏ a. **ROUTER> SHOW RUNNING-CONFIG**

❏ b. **ROUTER# SH RUN**

❏ c. **ROUTER>SHOW RUNNING-CONFIG**

❏ d. **ROUTER#SHOW RUNNING CONFIG**

❏ e. **ROUTER#SHOW RUNNING-CONFIG**

❏ f. **ROUTER>SH RUN**

❏ g. **ROUTER#DISPLAY RUNNING**

Question 59

Which of the following best describes the use of the enable password in a Cisco router?

○ a. The enable password is used to control access to the Ethernet port on the router.

○ b. The enable password is used to control remote Telnet access to a router.

○ c. The enable password is used to control access to any auxiliary ports the router may have.

○ d. The enable password is used to control access to the privileged mode of the router.

○ e. The enable password is used to control access to the console port of the router.

○ f. The enable password is used to control access to the user mode of the router.

Question 60

Two common protocols used when encapsulating data over serial links are synchronous data-link control (SDLC) and high-level data-link control (HDLC). Which of the following best describes the use of HDLC?

○ a. HDLC is a network layer that is the Cisco default protocol for all serial (WAN) links.

○ b. HDLC is a transport layer protocol used by IBM networks when communicating over WAN links.

○ c. HDLC is a data-link layer protocol that is the Cisco default protocol for all serial (WAN) links.

○ d. HDLC is a presentation layer protocol used by IBM networks when communicating over WAN links.

Question 61

IGRP is a dynamic routing protocol used to automatically update entries in a routing table. Which of the following statements are true about IGRP? [Choose the three best answers]

❑ a. IGRP is a distance vector protocol that is proprietary to Cisco.

❑ b. To select the best route, IGRP only uses hop count.

❑ c. IGRP considers hop count, delay, bandwidth, and reliability when determining the best path.

❑ d. The maximum hop count used in IGRP is 15.

❑ e. The maximum hop count used in IGRP is 255.

❑ f. IGRP is a link state routing protocol.

Question 62

Which of the following IPX SAP access lists will prevent a Cisco router from passing print service advertisements sent from all Netware servers?

○ a. **ACCESS LIST 1000 DENY −1 7**

○ b. **ACCESS-LIST 1000 DENY −1 7**

○ c. **ACCESS-LIST 1000 DENY −1 4**

○ d. **ACCESSLIST 1000 DENY −1 4**

Question 63

Which of the following best describes the difference between a switched virtual circuit (SVC) and a permanent virtual circuit (PVC) as they are used in X.25 packet switching networks? [Choose the two best answers]

❑ a. A PVC is a temporary X.25 connection that is established on an as-needed basis and is closed when not in use.

❑ b. An SVC is a temporary X.25 connection that is established on an as-needed basis and is closed when not in use.

❑ c. An SVC is a permanent X.25 connection that is always available for use.

❑ d. A PVC is a permanent X.25 connection that is always available for use.

Question 64

ISDN, or Integrated Services Digital Network, is a communications standard that uses digital telephone lines to transmit voice, data, and video. ISDN is a dial-up service that is used on demand. ISDN protocol standards are defined by the International Telecommunication Union (ITU). Which of the following statements is true about ISDN protocol standards?

○ a. Protocols that begin with the letter *Q* define ISDN standards on the existing telephone network.

○ b. Protocols that begin with the letter *I* define ISDN concepts, interfaces, and terminology.

○ c. Protocols that begin with the letter *I* specify ISDN switching and signaling standards.

○ d. Protocols that begin with the letter *E* specify ISDN concepts, interfaces, and terminology.

Question 65

You manage a network that uses TCP/IP as the only protocol and is divided into multiple subnets. The IP configuration of one of your subnets is as follows:

```
Broadcast Address =
129.23.98.31 on interface E0
Subnet Mask = 255.255.255.248
```

Based on this information, what are the valid host IDs for the nodes in this subnet?

○ a. 129.23.98.32 to 129.23.98.63

○ b. 129.23.98.24 to 129.23.98.31

○ c. 129.23.98.26 to 129.23.98.63

○ d. 129.23.98.25 to 129.23.98.30

Question 66

Which of the following core components of a Cisco router is used by Cisco routers to store the startup or running configuration?

○ a. Flash

○ b. RAM

○ c. ROM

○ d. NVRAM

Question 67

What will a Cisco router do if during startup no configuration file is found in NVRAM?

○ a. The router will crash.

○ b. The router will look for the configuration file in ROM.

○ c. The router will look for the configuration file in Flash.

○ d. The router will automatically initiate the setup program that provides the dialog necessary to create a configuration file.

Question 68

Which of the following commands can be used to log out of a router from privileged mode? [Choose the four best answers]

❏ a. **LOGOUT**

❏ b. **LOGOFF**

❏ c. **DISABLE**

❏ d. **QUIT**

❏ e. **EXIT**

❏ f. **LEAVE-OFF**

❏ g. **QUIT-ROUTER [ID]**

Question 69

Which of the following is the correct key(s) used to move the cursor back one word in a Cisco router?

- ○ a. Ctrl+A
- ○ b. Ctrl+E
- ○ c. Ctrl+P
- ○ d. Ctrl+N
- ○ e. Esc+B
- ○ f. Tab
- ○ g. PEPSI

Question 70

Which of the following best describes the use of the console password in a Cisco router?

- ○ a. The console password is used to control access to the Ethernet port on the router.
- ○ b. The console password is used to control remote Telnet access to a router.
- ○ c. The console password is used to control access to any auxiliary ports the router may have.
- ○ d. The console password is used to control access to the privileged mode of the router.
- ○ e. The console password is used to control access to the console port of the router.
- ○ f. The console password is used to control access to the user mode of the router.

Answer Key #3

1. c	19. a	37. a	55. b
2. a, d	20. a, d, f	38. a, c, e, f	56. b
3. c	21. g	39. a	57. b, g
4. b	22. f	40. a	58. b, e
5. a	23. c	41. b	59. d
6. c	24. f	42. c	60. c
7. c	25. c	43. c	61. a, c, e
8. d	26. b	44. c	62. b
9. a	27. a, d, e, g	45. d	63. b, d
10. b	28. a	46. g	64. b
11. c	29. a	47. a, b, d	65. d
12. b	30. b, c, d	48. a	66. d
13. d	31. d	49. b	67. d
14. a	32. a, d	50. b	68. a, c, d, e
15. a	33. d	51. a	69. e
16. b	34. d	52. b, c, d	70. e
17. a	35. b	53. a, b	
18. c	36. b	54. b	

Question 1

The correct answer is c. The access list **ACCESS-LIST 110 PERMIT TCP HOST 77.90.50.22 HOST 210.34.5.20 EQ 25** will allow SMTP traffic from host **77.90.50.22** to host **210.34.5.20**. Answers a and d are incorrect, because the correct syntax for an extended IP access list specifies the source first, then the destination. Answer b is incorrect, because FTP uses port 21, and SMTP uses port 25. The syntax for creating an extended IP access list is as follows:

```
ACCESS-LIST [access-list-number] [permit|deny]
 [protocol] [source]
 [wildcard mask] [destination] [wildmask]
 [operator][port]
```

➤ *ACCESS-LIST*—This is the command to denote an access list.

➤ *[access-list-number]*—This can be a number between 100 and 199 and is used to denote an extended IP access list.

➤ *[permit|deny]*—This will allow (permit) or disallow (deny) traffic specified in this access list.

➤ *[protocol]*—This is used to specify the protocol that will be filtered in this access list. Common values are TCP, UDP, and IP (use of IP will denote all IP protocols).

➤ *[source]*—This specifies the source IP addressing information.

➤ *[wildcard mask]*—This can be optionally applied to further define the source. A wildmask can be used to control access to an entire IP network ID rather than a single IP address, and it will use the number 255 to mean "any" and the number 0 to mean "must match exactly."

The terms *host* and *any* may also be used here to more quickly specify a single host or an entire network. The **HOST** keyword is the same as the wildcard mask 0.0.0.0; the **ANY** keyword is the same as the wildcard mask 255.255.255.255.

➤ *[destination]*—This specifies the destination IP addressing information.

➤ *[wildmask]*—This can be optionally applied to further define the destination IP addressing.

➤ *[operator]*—This can be optionally applied to define how to interpret the value entered in the **[port]** section of the access list. Common values are equal to the port specified (**EQ**), less than the port number specified (**LT**), and greater than the port number specified (**GT**).

➤ *[port]*—This specifies the port number this access list will act on. Common port numbers include 21 (FTP), 23 (TELNET), 25 (SMTP), 53 (DNS), 69 (TFTP), and 80 (HTTP).

For more information, see Chapter 4 of *CCNA Routing and Switching Exam Cram, Second Edition*.

Question 2

The correct answers are a and d. Network layer protocols include Internet Protocol (IP, part of the TCP/IP protocol suite) and Internetwork Packet Exchange (IPX, part of the IPX/SPX protocol suite). None of the other answers are network layer protocols. Transport layer protocols include Transmission Control Protocol (TCP, part of the TCP/IP protocol suite) and Sequenced Packet Exchange (SPX, part of the IPX/SPX protocol suite). Session layer protocols include Network File System (NFS, used by SUN Microsystems and Unix with TCP/IP), structured query language (SQL, used to define database information requests), remote procedure calls (RPC, used in Microsoft network communication), X Window (used by Unix terminals), AppleTalk Session Protocol (ASP, used by Apple computers), and Digital Network Architecture Session Control Protocol (DNA SCP, used by IBM). Simple Mail Transfer Protocol (SMTP) and Telnet are application layer protocols. PICTure (PICT, Apple computer picture format) and Tagged Image File Format (TIFF) are presentation layer protocols.

For more information, see Chapter 4 of *CCNA Routing and Switching Exam Cram, Second Edition*.

Question 3

The correct answer is c. The router is in global configuration mode. The **CONFIG T** command is used in privileged mode to access global configuration mode. Global configuration mode is displayed at the router prompt with "config" in parentheses **(RouterA(config)#)**. To exit global configuration mode, use the keys Control and Z (Ctrl+Z). Answer a is incorrect, because in user mode, the router prompt is followed by an angle bracket **(ROUTER>)**. User mode is used to display basic router system information and to connect to remote devices. Answer b is incorrect, because in privileged mode, the router prompt is followed by a pound sign **(ROUTER#)**. Privileged mode is used to modify and view the configuration of the router, as well as to set IOS parameters. Privileged mode also contains the **CONFIGURE** command, which is used to access other configuration modes, such as global configuration mode and interface mode. To access privileged mode, enter the **ENABLE** command from user mode. Answer d is incorrect because there is sufficient information to answer this question.

For more information, see Chapter 8 of *CCNA Routing and Switching Exam Cram, Second Edition*.

Question 4

The correct answer is b. Typing a command followed by a space, then a question mark will display any keywords that can be used with that command. Typing "CLOCK ?" at the router prompt will return **SET**, the keyword used with the **CLOCK** command. Answer a is incorrect, because no space is between "CLOCK" and the question mark. Typing "CLOCK?" will result in the router displaying **CLOCK**, the command that begins with those letters. Answer c is incorrect, because **HELP CLOCK** is not a valid Cisco IOS command. Answer d is incorrect, because **SHOW CLOCK** will display the current time but will not display the keywords used with the **CLOCK** command.

The Cisco IOS includes a context-sensitive help feature that can assist an operator when entering commands into a Cisco router. Context-sensitive help provides information on available IOS commands, command line syntax, and available command keywords. Accessing the features of context-sensitive help involves the correct use of the question mark key (?). The following list details the common use of the question mark (?) key in context-sensitive help:

➤ *ROUTER#?*—Typing a single question mark at the router prompt will list all available commands for that particular router mode.

➤ *ROUTER#<COMMAND> ?*—Typing a command followed by a space, then a question mark will display any keywords that can be used with that command. For example, typing "CLOCK ?" at the router prompt will return **SET**, the keyword used with the **CLOCK** command.

➤ *ROUTER#<PARTIAL COMMAND>?*—Typing a partial command followed *immediately* by a question mark (no space between partial command and question mark) will display a list of commands that begin with the partial command entry. For example, typing "CL?" at the router prompt will return **CLEAR** and **CLOCK**, the two IOS commands that begin with "CL." Note that if a space is included between a partial or complete command and a question mark, help will interpret this as a keyword request and could result in an "ambiguous command" message. For example, typing "CL ?" (CL, a space, then a question mark) will result in the message "Ambiguous command: "cl " ", because two commands begin with "CL:" (**CLOCK** and **CLEAR**).

For more information, see Chapter 8 of *CCNA Routing and Switching Exam Cram, Second Edition.*

Question 5

The correct answer is a. Typing the **SHOW HISTORY** command will display all of the commands currently stored in the history buffer. All of the other answers are not valid Cisco IOS commands, therefore, answers b, c, and d are incorrect. The Cisco IOS includes a command history feature that can be used to store previously entered commands in a history buffer. Commands in the history buffer can be recalled by an operator at the router prompt, using Ctrl+P and C+N, saving configuration time. The command history feature is enabled by default, but it can be disabled, and the size of the history buffer can be modified. The following list details the common commands used when modifying the command history:

➤ *SHOW HISTORY*—Typing the **SHOW HISTORY** command will display all of the commands currently stored in the history buffer. Pressing Ctrl+P will recall previous commands in the command history at the router prompt (the up arrow will do this as well). Pressing Ctrl+N will return to more recent commands in the command history after recalling previous commands (the down arrow will do this as well).

➤ *TERMINAL HISTORY SIZE (0-256)*—Typing **TERMINAL HISTORY SIZE (0-256)** will allow you to set the number of command lines the command history buffer will hold. You must enter a number between 0 and 256. (The default is 10 commands.)

➤ *TERMINAL NO EDITING*—Typing **TERMINAL NO EDITING** will disable the command history feature. It is enabled by default.

➤ *TERMINAL EDITING*—Typing **TERMINAL EDITING** will enable the command history feature if it has been disabled.

For more information, see Chapter 8 of *CCNA Routing and Switching Exam Cram, Second Edition.*

Question 6

The correct answer is c. The **SHOW STARTUP-CONFIG** command can be used in privileged mode to display the startup configuration file stored in NVRAM. Answers a and d are incorrect, because the correct command was issued (**SHOW STARTUP-CONFIG**). Answer b is incorrect, because the router is in user mode, not privileged mode. In user mode, the router prompt is followed by an angle bracket (**ROUTER>**). The **SHOW STARTUP-CONFIG** command cannot be used in either user or global configuration modes. In privileged mode, the router prompt is followed by a pound sign (**ROUTER#**).

For more information, see Chapter 8 of *CCNA Routing and Switching Exam Cram, Second Edition.*

Question 7

The correct answer is c. The **SHOW IP ACCESS-LIST 127** command will display the lines in IP access list 127 that were matched by the traffic that passed through the router. Answer a is incorrect, because no hyphen is between **ACCESS** and **LIST**. Answers b and d are incorrect, because **SHOW IP INTERFACE** will not display any matches to the individual lines of the access list like the **SHOW ACCESS-LIST** commands. Answer a is incorrect, because the correct syntax for the **SHOW IP ACCESS-LIST 127** command has a hyphen between **ACCESS** and **LIST**. Access lists can be used to filter the traffic that is handled by a Cisco router. Several Cisco IOS commands can be used to view the various access lists in use on a Cisco router. The following details the most common Cisco IOS commands used to view these lists:

➤ *SHOW ACCESS-LISTS*—This command will display all of the access lists in use on the router, and it will also show each line of the access list and return the number of times a packet matched that line. This command can also be used with a specific access list number to display this detail on a single IP access list. *SHOW ACCESS-LISTS* will not display what interface an access list has been applied to, as the **SHOW INTERFACE** commands do.

➤ *SHOW IP ACCESS-LIST*—This will display all IP access lists in use on the router. It will also show each line of the access list and return the number of times a packet matched that line. This command can also be used with a specific IP access list number to display this detail on a single IP access list. **SHOW IP ACCESS-LIST** will not display what interface an access list has been applied to, as the **SHOW INTERFACE** commands do.

➤ *SHOW IPX ACCESS-LIST*—This will display all IP access lists in use on the router. It will also show each line of the access list and return the number of times a packet matched that line. This command can also be used with a specific IPX access list number to display this detail on a single IPX access list. **SHOW IPX ACCESS-LIST** will not display what interface an access list has been applied to, as the **SHOW INTERFACE** commands do.

➤ *SHOW IP INTERFACE [interface number]*—This will display the IP access lists that have been applied to a specific interface. It will not display any matches to the individual lines of the access list, as the **SHOW ACCESS-LIST** commands do.

➤ *SHOW IPX INTERFACE [interface number]*—This will display the IP access lists that have been applied to a specific interface. It will not display any matches to the individual lines of the access list, as the **SHOW ACCESS-LIST** commands do.

For more information, see Chapter 13 of *CCNA Routing and Switching Exam Cram, Second Edition.*

Question 8

The correct answer is d. The network ID of this address is 2c, and the node ID of this address is 1234.1111.2222. All of the other answers are incorrect, because they are incorrect combinations of the network ID. The IPX/SPX protocol is used in Novell NetWare networks. An IPX address is a network layer, hierarchical address that is similar in format to an IP address in that IPX addresses contain a network ID and a node ID. The network ID of an IPX address identifies the network that an IPX node is a part of and will be the same for all IPX computers in the same network. The node ID of an IPX address is unique to each host on an IPX network and is assigned dynamically using the MAC address of the computer. This makes configuring IPX/SPX addresses much easier than TCP/IP addresses. The network ID of an IPX address is assigned by an administrator and can be up to 32 bits in length. The network ID is usually expressed in decimal notation without the leading zeros. The node ID of an IPX address is 48 bits in length and can be expressed in decimal notation by four digits separated by three periods. A complete IPX address is expressed in decimal notation by combining the network ID with the node ID. For example, if an IPX host has a node ID of 1234.1234.1234 and is part of IPX network 2c, the IPX address for that host would be 2c.1234.1234.1234.

For more information, see Chapter 12 of *CCNA Routing and Switching Exam Cram, Second Edition.*

Question 9

The correct answer is a. **SHOW IP PROTOCOL** will display the routing protocols in use for IP on the entire router, as well as update frequency and filter information. Answers b and d are incorrect, because the **SHOW IP ROUTE** command is used to display the contents of the routing table and does not display the routing protocols in use for IP on the entire router. Answer c is incorrect, because the **SHOW IP PROTOCOL** command does not have a hyphen between **IP** and **PROTOCOL**.

For more information, see Chapter 8 of *CCNA Routing and Switching Exam Cram, Second Edition.*

Question 10

The correct answer is b. The OSI model defines standards used in networking and comprises a seven-layer model. The presentation layer of the OSI model is primarily used as a "translator" for the application layer and uses Abstract Syntax Notation One (ASN 1) to negotiate or translate information to the application layer. Since no other layer of the OSI model performs this function, all of the other answers are incorrect. The following list outlines the seven layers of the OSI model and their functions:

➤ *Application*—The "window" to networking used by programs, the application layer is responsible for:

  ➤ Verifying that the appropriate resources are present to initiate a connection with the destination node.

  ➤ Verifying the identity of the destination node.

Application layer protocols include Telnet, File Transfer Protocol (FTP), Simple Mail Transfer Protocol (SMTP), World Wide Web (WWW), Electronic Data Interchange (EDI), and Wide Area Information Server (WAIS).

➤ *Presentation*—Essentially a translator, the presentation layer is responsible for:

  ➤ Translating text and data syntax, such as Extended Binary-Coded Decimal Interchange Code (EBCDIC, used in IBM systems) to American Standard Code for Information Interchange (ASCII, used in PCs and most computer systems).

  ➤ Using Abstract Syntax Notation One (ASN 1) to perform data translation.

Presentation layer protocols include PICTure (PICT, Apple computer picture format), Tagged Image File Format (TIFF), Joint Photographic Experts Group (JPEG), Musical Instrument Digital Interface (MIDI), Motion Picture Expert Group (MPEG), and Quick Time (audio/video application).

➤ *Session*—Used to coordinate communication between nodes, the session layer is responsible for:

➤ Creating, maintaining, and ending a communication session.

➤ Coordinating service requests and responses that occur between nodes across the network.

Session layer protocols include Network File System (NFS, used by SUN Microsystems and Unix with TCP/IP), structured query language (SQL, used to define database information requests), remote procedure calls (RPCs, used in Microsoft network communication), X Window (used by Unix terminals), AppleTalk Session Protocol (ASP, used by Apple computers), and Digital Network Architecture Session Control Protocol (DNA SCP, used by IBM).

➤ *Transport*—Reliable end-to-end communication is the primary function of the transport layer. Its many responsibilities include:

➤ Ensuring flow control, so the amount of data being sent from one node will not overwhelm the destination node.

➤ Ensuring the ability of multiple connections to utilize a single transport (multiplexing).

➤ Ensuring the reliable transfer of data. The transport layer implements connection-oriented services between nodes, which utilize a three-way handshake (synchronization, acknowledgment, and data-transfer) to efficiently transfer data.

➤ Ensuring positive acknowledgment of the process of a node waiting for an acknowledgment from the destination node before sending data.

➤ Windowing, a form of flow control that specifies how much data will be transferred between acknowledgments.

Transport layer protocols include Transmission Control Protocol (TCP, part of the TCP/IP protocol suite) and Sequenced Packet Exchange (SPX, part of the IPX/SPX protocol suite).

➤ *Network*—Selecting the appropriate path a packet should take to get to its intended destination is the function of the network layer. It is responsible for routing, which is the process of using a network layer address to determine the best path a packet will travel to its destination. Network layer protocols include Internet Protocol (IP, part of the TCP/IP protocol suite) and Internetwork Packet Exchange (IPX, part of the IPX/SPX protocol suite).

➤ *Data Link*—Divided into two sublayers (Media Access Control [MAC] and Logical Link Control [LLC]), the data-link layer is responsible for:

➤ Preparing data from upper layers to be transmitted over the physical medium by encapsulating upper-layer data into frames. This frame includes the source and destination of MAC addresses in the frame header.

➤ Converting data into bits, so it can be transmitted by the physical layer.

➤ Adding a cyclical redundancy check (CRC) to the end of a frame, which is used for error checking at the data-link layer.

Data-link protocols include High-Level Data Link Control (HDLC, the Cisco default encapsulation for serial connections), Synchronous Data Link Control, (SDLC, used in IBM networks), Link Access Procedure Balanced (LAPB, used with X.25), X.25 (packet switching network), Serial Line Internet Protocol (SLIP, an older TCP/IP dial-up protocol), Point-to-Point Protocol (PPP, a newer dial-up protocol), Integrated Services Digital Network (ISDN, a dial-up digital service), and frame relay (a packet switching network).

➤ *Physical*—The lowest layer of the OSI model, the physical layer defines the electrical functionality required to send and receive bits over a given physical medium. Specifications that define the voltage levels and physical components of a network are defined at the physical layer. Protocols are not specified at the physical layer, because they are implemented as software. Examples of the standards for sending data over the physical medium are Ethernet (the most widely used standard in networking), Token Ring (IBM's proprietary networking topology), and Fiber Distributed Data Interface (FDDI, a standard for fiber optic networks, commonly used as a backbone).

For more information, see Chapter 4 of *CCNA Routing and Switching Exam Cram, Second Edition.*

Question 11

The correct answer is c. **CONFIGURE OVERWRITE-NETWORK (CONFIG O)** is used to copy a configuration file into NVRAM from a TFTP server. Answer a is incorrect, because the **CONFIGURE TERMINAL** command is used to enter configuration commands into the router from the console port or through Telnet. Answer b is incorrect, because the **CONFIGURE OVERWRITE NETWORK** command is used to copy a configuration file into NVRAM from a TFTP server. Answer d is incorrect, because the **CONFIGURE MEMORY** command is used to execute the configuration stored in NVRAM. The **CONFIGURE** command, executed in privileged mode, is used to enter the configuration mode of a Cisco router. The four parameters used with the **CONFIGURE** command are **TERMINAL, MEMORY, OVERWRITE-NETWORK,** and **NETWORK.** The following list outlines the functions of these parameters:

➤ *CONFIGURE TERMINAL (CONFIG T)*—This command is used to enter configuration commands into the router from the console port or through Telnet.

➤ *CONFIGURE NETWORK (CONFIG NET)*—This command is used to copy the configuration file from a TFTP server into the router's RAM.

➤ *CONFIGURE OVERWRITE-NETWORK (CONFIG O)*—This command is used to copy a configuration file into NVRAM from a TFTP server. Use of the **CONFIGURE OVERWRITE-NETWORK** will not alter the running configuration.

➤ *CONFIGURE MEMORY (CONFIG MEM)*—This command is used to execute the configuration stored in NVRAM. It will copy the startup configuration to the running configuration.

For more information, see Chapter 8 of *CCNA Routing and Switching Exam Cram, Second Edition.*

Question 12

The correct answer is b. Pressing the Control key along with the letter *P* will recall previous commands in the command history. Since this is the only use for Ctrl+P, all of the other answers are incorrect. Configuring a Cisco router can sometimes involve long, detailed command lines that can be slow to navigate. For this reason, the Cisco IOS includes a number of shortcuts through them. The following list details some of the more common methods used to navigate the command line:

➤ *Ctrl+A*—Pressing the Control key along with the letter *A* will move the cursor to the beginning of the command line.

➤ *Ctrl+E*—Pressing the Control key along with the letter *E* will move the cursor to the end of the command line.

➤ *Ctrl+P*—Pressing the Control key along with the letter *P* will recall previous commands in the command history. (The up arrow will do this as well.)

➤ *Ctrl+N*—Pressing the Control key along with the letter *N* will return to more recent commands in the command history after recalling previous commands. (The down arrow will do this as well.)

➤ *Esc+B*—Pressing the Escape key along with the letter *B* will move the cursor back one word.

➤ *Esc+F*—Pressing the Escape key along with the letter *F* will move the cursor forward one word.

➤ *Left and Right Arrow Keys*—These arrow keys will move the cursor one character left and right, respectively.

➤ *Tab*—Pressing the Tab key will complete an entry typed at the router prompt.

For more information, see Chapter 8 of *CCNA Routing and Switching Exam Cram, Second Edition.*

Question 13

The correct answer is d. The **ENABLE SECRET PENNICK** command will set the password to **PENNICK** and encrypt this password when it is entered. Answer a is incorrect, because **ENABLE PASSWORD PENNICK** will set an enable password, but it will not be encrypted. Answer b is incorrect, because **ENABLE SECRET PASSWORD PENNICK** is not a valid Cisco IOS command. Answer c is incorrect, because **LINE ENABLE SECRET PENNICK** is not a valid Cisco IOS command. Both the enable secret password and the enable password can be used to control access to the privileged mode of the router. The difference between enable secret and enable password is the enable secret password will be encrypted for additional security. The enable secret password will take precedence over the enable password if both are enabled. To configure the enable secret password, use the **ENABLE SECRET [password]** command. If an enable password is specified, a password must be entered after the **ENABLE** command to successfully access privileged mode. To configure the enable password, use the **ENABLE PASSWORD [*password*]** command.

For more information, see Chapter 8 of *CCNA Routing and Switching Exam Cram, Second Edition.*

Question 14

The correct answer is a. Because the subnet mask 255.255.255.192 uses 26 bits, and an IP address is a total of 32 bits, 6 bits remain to create host IDs on this network. An IP address is a 32-bit network layer address that comprises a network ID and a host ID. A subnet mask is used to distinguish the network portion of an IP address from the host portion. The 32 bits in an IP address are expressed in dotted decimal notation in four octets (four 8-bit numbers separated by periods, that is, 208.100.50.90). The subnet mask 255.255.255.192 is said to use 26 bits, because it takes 26 of 32-bit positions to create. 255.255.255.192 can be expressed in binary as 11111111.11111111.11111111.11000000, which is three octets using 8 bits each (24) and 2 bits used in the fourth octet to make the number 192. Because the subnet mask is using 26 bits, and an IP address is a total of 32 bits, 6 bits remain to create host IDs on this network. Because of this, all of the other answers are invalid.

For more information, see Chapter 8 of *CCNA Routing and Switching Exam Cram, Second Edition.*

Question 15

The correct answer is a. The OSI model defines a layered approach to network communication. Each layer of the OSI model adds its own layer-specific information and passes it on to the next layer, until it leaves the computer and goes out onto the network. This process is known as encapsulation, and it involves five basic steps:

1. User information is converted to data.

2. Data is converted to segments.

3. Segments are converted to packets or datagrams.

4. Packets and datagrams are converted to frames.

5. Frames are converted to bits.

For more information, see Chapter 3 of *CCNA Routing and Switching Exam Cram, Second Edition.*

Question 16

The correct answer is b. The **SHOW IPX ACCESS-LIST** command will display all IPX access lists in use on the router. Answer a is incorrect, because no hyphen is between **ACCESS** and **LIST**. Answer c is incorrect, because of the illegal hyphen between **SHOW** and **IPX**. Answer d is incorrect, because the **SHOW IP ACCESS-LIST** command cannot be followed with a range of numbers. Access lists can be used to filter the traffic that is handled by a Cisco router. Several Cisco IOS commands can be used to view the various access lists in use on a Cisco router. The following list details the most common Cisco IOS commands used to view access lists:

➤ *SHOW ACCESS-LISTS*—This command will display all of the access lists in use on the router. It will also show each line of the access list and return the number of times a packet matched that line. This command can also be used with a specific access list number to display this detail on a single IP access list. **SHOW ACCESS-LISTS** will not display what interface an access list has been applied to, as the **SHOW INTER-FACE** commands do.

➤ *SHOW IP ACCESS-LIST*—This command will display all IP access lists in use on the router. It will also show each line of the access list and return the number of times a packet matched that line. This command can also be used with a specific IP access list number to display this detail on a single IP access list. **SHOW IP ACCESS-LISTS** will not display what interface an access list has been applied to, as the **SHOW INTERFACE** commands do.

➤ *SHOW IPX ACCESS-LIST*—This command will display all IP access lists in use on the router. It will also show each line of the access list and return the number of times a packet matched that line. This command can also be used with a specific IPX access list number to display this detail on a single IPX access list. **SHOW IPX ACCESS-LISTS** will not display what interface an access list has been applied to, as the **SHOW INTERFACE** commands do.

➤ *SHOW IP INTERFACE [interface number]*—This command will display the IP access lists that have been applied to a specific interface. It will not display any matches to the individual lines of the access list, as the **SHOW ACCESS-LIST** commands do.

➤ *SHOW IPX INTERFACE [interface number]*—This will display the IP access lists that have been applied to a specific interface. It will not display any matches to the individual lines of the access list, as the **SHOW ACCESS-LIST** commands do.

For more information, see Chapter 13 of *CCNA Routing and Switching Exam Cram, Second Edition.*

Question 17

The correct answer is a. The **IP ACCESS-GROUP 10 OUT** is the correct command to use in this question to apply the access list 10 to the outgoing interface of the router. Answer b is incorrect, because the correct syntax for the **ACCESS GROUP** command uses a hyphen between **ACCESS** and **GROUP**. Answer b is incorrect, because the correct syntax for the **IP ACCESS-GROUP** command begins with IP. Answer d is incorrect, because the correct syntax for the **IP ACCESS-GROUP** command specifies the access list number after IP

ACCESS-GROUP, not the interface number. For an IP access list to filter traffic, it must be applied to an interface. This is done using the **IP ACCESS-GROUP** command, which must be entered in interface configuration mode. To access this mode, use the **INTERFACE [interface-number]** command from global configuration mode. When the router is in interface configuration mode, the router prompt displays "config-if" in parentheses, followed by a pound sign ((ROUTER(config-if)#). The **IP ACCESS GROUP** command has the following syntax:

```
IP ACCESS-GROUP [access-list-number] [out|in]
```

➤ *[out|in]*—This parameter specifies where the access list will be applied on the router. **OUT** will cause the router to apply the access list to all outgoing packets, and **IN** will cause the router to apply the access list to all incoming packets.

Access lists can be used to filter the traffic that is handled by a Cisco router. An IP access list will filter IP traffic and can be created as a standard IP access list or an extended IP access list. A standard IP access list can be used to permit or deny access based on IP addressing information only and can only act on the source IP addressing information. The syntax for creating a standard IP access list is:

```
ACCESS-LIST [access-list-number] [permit|deny] source
[wildcard mask]
```

➤ *ACCESS-LIST*—This command is followed by the **access list number**. When using a standard IP access list, this can be a number between 1 and 99.

➤ *[permit|deny]*—This statement will either allow or disallow traffic in this access list.

➤ *source*—This statement is used to specify the source IP addressing information. A standard IP access list can only act on a source address.

➤ *[wildcard mask]*—This is an optional parameter in a standard IP access list. If no **wildcard mask** is specified, the access list will only act on a single IP address that is specified in the **source**. A wildmask can be used to control access to an entire IP network ID rather than a single IP address. A **wildcard mask** will use the number 255 to mean "any" and the number 0 to mean "must match exactly."

For more information, see Chapter 13 of *CCNA Routing and Switching Exam Cram, Second Edition.*

Question 18

The correct answer is c. The **IP ROUTE 131.200.0.0 255.255.0.0 220.200.200.84** command will create a static route in the router's routing table that will send all packets destined for network **131.200.0.0 255.255.0.0** to router **220.200.200.84**. Answer a is incorrect, because the **IP ROUTE** command does not have a hyphen between **IP** and **ROUTE**. Answer b is incorrect, because **STATIC ROUTE** is not a valid Cisco IOS command. Answer d is incorrect, because the **IP ROUTE** command specifies the source network first, then the destination router. A routing table can be of two basic types: static and dynamic. A static routing table is one that has its entries entered manually by an operator. On a Cisco router, a static route can be entered using the **IP ROUTE** command. The syntax for the **IP ROUTE** command is as follows:

```
IP ROUTE [network] [subnet mask] [IP address] [distance]
```

➤ *IP ROUTE*—This command signifies this will be a static entry in the routing table.

➤ *[network]*—This is the destination network ID.

➤ *[subnet mask]*—This is the subnet mask of the destination network.

➤ *[IP address]*—This is the IP address of the router that will receive all packets that have the address specified in the **NETWORK** statement.

➤ *[distance]*—An optional parameter, this is the administrative distance of this route, which tells the router the relative importance of this route. If a router has multiple entries in its routing table for the same network ID, it will use the route with the lowest administrative distance. The default administrative distance for a static route is 1. (The number 1 is the second highest priority for administrative distance; 0 is the highest and signifies a directly connected interface.)

For more information, see Chapter 8 of *CCNA Routing and Switching Exam Cram, Second Edition.*

Question 19

The correct answer is a. Maximum hop count involves setting a maximum hop count that a packet can travel before it is discarded. Answer b is incorrect, because route poisoning occurs when a router poisons a route by setting a hop count of 16. Answer c is incorrect, because split horizon states that a packet cannot be sent back in the direction from which it was received. Answer d is incorrect, because hold-down is a specified amount of time a router will wait before updating its routing table. A routing loop occurs when a link on a router fails, and some routers do not receive the updated information. The routers that do not have the updated information in their routing table will advertise to other routers that they have a route to the failed link and will begin receiving the packets destined for that network. Because this route is failed and because there is no real way to route the packet to its destination, packets will end up getting passed back and forth between routers and never reach their destination. To combat the problem of routing loops, distance vector routing protocols utilize four basic techniques: maximum hop count, split horizon, route poisoning, and hold-downs. The following list describes the function of each:

➤ *Maximum Hop Count*—This function sets a maximum hop count that a packet can travel before it is discarded. RIP uses a maximum hop count of 15, so in the event of a routing loop, a packet that travels more than 15 hops will be discarded.

➤ *Split Horizon*—This function states that a packet cannot be sent back in the direction from which it was received. This prevents the routing loop of two routers passing a packet back and forth between each other.

➤ *Route Poisoning*—When a router that is using a distance vector routing protocol has a link that fails, it will "poison" its route by setting the hop count of the failed link to a number over the maximum hop count, so other routers will deem the route unreachable. In the case of RIP, a router can poison a route by setting a hop count of 16. (Fifteen is the maximum hop count used by RIP.)

➤ *Hold-Downs*—This function specifies the amount of time a router will wait before updating its routing table. Hold-downs prevent routing loops by preventing routers from updating new information too quickly, allowing time for a failed link to be re-established.

For more information, see Chapter 11 of *CCNA Routing and Switching Exam Cram, Second Edition*.

Question 20

The correct answers are a, d, and f. IP address 150.150.33.90 is a Class B address with a network ID of 150.150.0.0 and a host ID of 33.90. Answer b is incorrect, because this is a Class C address. Answer c is incorrect, because the network ID is 150.150.0.0. Answer e is incorrect, because the host ID is 33.90. An IP address is a 32-bit network layer address that comprises a network ID and a host ID. TCP/IP addresses are divided into three main classes: Class A,

Class B, and Class C. The "first octet" rule specifies that an IP address can be identified by class according to the number in the first octet. The address range for each address class is as follows:

➤ *Class A*—IP addresses with numbers from 1 to 126 in the first octet. (Note: 127 is not used as a network ID in IP, it is reserved for localhost loopback testing.)

➤ *Class B*—IP addresses with numbers from 128 to 191 in the first octet.

➤ *Class C*—IP addresses with numbers from 192 to 223 in the first octet.

When looking at any IP address, it is the number in the *first* octet that determines the address class of that address.

➤ The IP address 3.100.32.7 is a Class A address, because it *begins* with 3, a number between 1 and 126.

➤ The IP address 145.100.32.7 is a Class B address, because it *begins* with 145, a number between 128 and 191.

➤ The IP address 210.100.32.7 is a Class C address, because it *begins* with 210, a number between 192 and 223.

An IP address is made up of a network ID and a host ID. The network ID identifies the network a node is on, and the host ID identifies the individual node. An IP address is specified using four individual numbers separated by three periods. Each one of these "sections" is known as an octet (because it contains 8 bits). Each address class uses different combinations of octets to specify a network ID and host ID. The configuration is as follows:

➤ *Class A*—This class uses the first octet for the network ID and the last three octets for the host ID.

➤ IP address 2.3.45.9 is a Class A address (it begins with 2, a number between 1 and 126). The network ID of this IP address is 2, and the host ID is 3.45.9, because a Class A address uses the first octet for the network ID and the last three octets for the host ID.

➤ *Class B*—This class uses the first two octets for the network ID and the last 2 octets for the host ID.

➤ IP address 167.211.32.5 is a Class B address (it begins with 167, a number between 128 and 191). The network ID of this IP address is 167.211, and the host ID is 32.5, because a Class B address uses the first two octets for the network ID and the last two octets for the host ID.

➤ *Class C*—This class uses the first three octets for the network ID and the last octet for the host ID.

> ➤ IP address 200.3.11.75 is a Class C address (it begins with 200, a number between 192 and 223). The network ID of this IP address is 200.3.11 and the host ID is 75, because a Class C address uses the first three octets for the network ID and the last octet for the host ID.

For more information, see Chapter 8 of *CCNA Routing and Switching Exam Cram, Second Edition.*

Question 21

The correct answer is g. The only protocol that does not reside at the session layer is HDLC (the Cisco default encapsulation for serial connections), which is a data-link protocol. The session layer of the OSI model is used to coordinate communication between nodes and is also responsible for creating, maintaining, and ending a communication session. Session layer protocols include NFS, used by SUN Microsystems and Unix with TCP/IP; SQL, used to define database information requests; RPC, used in Microsoft network communication; X Window, used by Unix terminals; ASP, used by Apple computers; and DNA SCP, used by IBM.

For more information, see Chapter 4 of *CCNA Routing and Switching Exam Cram, Second Edition.*

Question 22

The correct answer is f. The OSI model defines standards used in networking and comprises a seven-layer model. The data-link layer of the OSI model is responsible for preparing data from upper layers to be transmitted over the physical medium by encapsulating upper-layer data into frames and adding a CRC to the end of a frame, used for error checking at the data-link layer. Since the data-link layer is the only layer of the OSI model that performs this function, all of the other answers are incorrect. The following list outlines the seven layers of the OSI model and their functions:

➤ *Application*—The "window" to networking used by programs, the application layer is responsible for:

> ➤ Verifying that the appropriate resources are present to initiate a connection with the destination node.

> ➤ Verifying the identity of the destination node.

Application layer protocols include Telnet, File Transfer Protocol (FTP), Simple Mail Transfer Protocol (SMTP), World Wide Web (WWW), Electronic Data Interchange (EDI), and Wide Area Information Server (WAIS).

➤ *Presentation*—Essentially a translator, the presentation layer is responsible for:

> ➤ Translating text and data syntax, such as Extended Binary-Coded Decimal Interchange Code (EBCDIC, used in IBM systems) to American Standard Code for Information Interchange (ASCII, used in PCs and most computer systems).

> ➤ Using Abstract Syntax Notation One (ASN 1) to perform data translation.

Presentation layer protocols include PICTure (PICT, Apple computer picture format), Tagged Image File Format (TIFF), Joint Photographic Experts Group (JPEG), Musical Instrument Digital Interface (MIDI), Motion Picture Expert Group (MPEG), and Quick Time (audio/video application).

➤ *Session*—Used to coordinate communication between nodes, the session layer is responsible for:

> ➤ Creating, maintaining, and ending a communication session.

> ➤ Coordinating service requests and responses that occur between nodes across the network.

Session layer protocols include Network File System (NFS, used by SUN Microsystems and Unix with TCP/IP), structured query language (SQL, used to define database information requests), remote procedure calls (RPCs, used in Microsoft network communication), X Window (used by Unix terminals), AppleTalk Session Protocol (ASP, used by Apple computers), and Digital Network Architecture Session Control Protocol (DNA SCP, used by IBM).

➤ *Transport*—Reliable end-to-end communication is the primary function of the transport layer. Its many responsibilities include:

> ➤ Ensuring flow control, so the amount of data being sent from one node will not overwhelm the destination node.

> ➤ Ensuring the ability of multiple connections to utilize a single transport (multiplexing).

> ➤ Ensuring the reliable transfer of data. The transport layer implements connection-oriented services between nodes, which utilize a three-way handshake (synchronization, acknowledgment, and data-transfer) to efficiently transfer data.

➤ Ensuring positive acknowledgment of the process of a node waiting for an acknowledgment from the destination node before sending data.

➤ Windowing, a form of flow control that specifies how much data will be transferred between acknowledgments.

Transport layer protocols include Transmission Control Protocol (TCP, part of the TCP/IP protocol suite) and Sequenced Packet Exchange (SPX, part of the IPX/SPX protocol suite).

➤ *Network*—Selecting the appropriate path a packet should take to get to its intended destination is the function of the network layer. It is responsible for routing, which is the process of using a network layer address to determine the best path a packet will travel to its destination. Network layer protocols include Internet Protocol (IP, part of the TCP/IP protocol suite) and Internetwork Packet Exchange (IPX, part of the IPX/SPX protocol suite).

➤ *Data Link*—Divided into two sublayers (Media Access Control [MAC] and Logical Link Control [LLC]), the data-link layer is responsible for:

➤ Preparing data from upper layers to be transmitted over the physical medium by encapsulating upper-layer data into frames. This frame includes the source and destination of MAC addresses in the frame header.

➤ Converting data into bits, so it can be transmitted by the physical layer.

➤ Adding a cyclical redundancy check (CRC) to the end of a frame, which is used for error checking at the data-link layer.

Data-link protocols include High-Level Data Link Control (HDLC, the Cisco default encapsulation for serial connections), Synchronous Data Link Control, (SDLC, used in IBM networks), Link Access Procedure Balanced (LAPB, used with X.25), X.25 (packet switching network), Serial Line Internet Protocol (SLIP, an older TCP/IP dial-up protocol), Point-to-Point Protocol (PPP, a newer dial-up protocol), Integrated Services Digital Network (ISDN, a dial-up digital service), and frame relay (a packet switching network).

➤ *Physical*—The lowest layer of the OSI model, the physical layer defines the electrical functionality required to send and receive bits over a given physical medium. Specifications that define the voltage levels and physical components of a network are defined at the physical layer. Protocols are not specified at the physical layer, because they are implemented as software. Examples of the standards for sending data over the physical medium are Ethernet (the most widely used standard in networking), Token Ring (IBM's proprietary networking topology), and Fiber Distributed Data Interface (FDDI, a standard for fiber optic networks, commonly used as a backbone).

For more information, see Chapter 4 of *CCNA Routing and Switching Exam Cram, Second Edition.*

Question 23

The correct answer is c. The range of numbers used to specify a standard IPX access list is 800–899. None of the other answers correctly describes the range used in a standard IPX access list. Access lists can be easily identified by referring to the number after the **ACCESS-LIST** command. The five major types of access lists utilize the following ranges of numbers:

➤ *IP Standard Access List*—1–99

➤ *IP Extended Access List*—100–199

➤ *IPX Standard Access List*—800–899

➤ *IPX Extended Access List*—900–999

➤ *IPX SAP Access List*—1000–1099

For more information, see Chapter 13 of *CCNA Routing and Switching Exam Cram, Second Edition.*

Question 24

The correct answer is f. The OSI model defines standards used in networking and comprises a seven-layer model. The data-link layer of the OSI model is responsible for preparing data from upper layers to be transmitted over the physical medium by encapsulating upper-layer data into frames. Since no other layer of the OSI model performs this function, all of the other answers are incorrect. The following list outlines the seven layers of the OSI model and their functions:

➤ *Application*—The "window" to networking used by programs, the application layer is responsible for:

  ➤ Verifying that the appropriate resources are present to initiate a connection with the destination node.

  ➤ Verifying the identity of the destination node.

Application layer protocols include Telnet, File Transfer Protocol (FTP), Simple Mail Transfer Protocol (SMTP), World Wide Web (WWW), Electronic Data Interchange (EDI), and Wide Area Information Server (WAIS).

➤ *Presentation*—Essentially a translator, the presentation layer is responsible for:

  ➤ Translating text and data syntax, such as Extended Binary-Coded Decimal Interchange Code (EBCDIC, used in IBM systems) to American Standard Code for Information Interchange (ASCII, used in PCs and most computer systems).

  ➤ Using Abstract Syntax Notation One (ASN 1) to perform data translation.

Presentation layer protocols include PICTure (PICT, Apple computer picture format), Tagged Image File Format (TIFF), Joint Photographic Experts Group (JPEG), Musical Instrument Digital Interface (MIDI), Motion Picture Expert Group (MPEG), and Quick Time (audio/video application).

➤ *Session*—Used to coordinate communication between nodes, the session layer is responsible for:

  ➤ Creating, maintaining, and ending a communication session.

  ➤ Coordinating service requests and responses that occur between nodes across the network.

Session layer protocols include Network File System (NFS, used by SUN Microsystems and Unix with TCP/IP), structured query language (SQL, used to define database information requests), remote procedure calls (RPCs, used in Microsoft network communication), X Window (used by Unix terminals), AppleTalk Session Protocol (ASP, used by Apple computers), and Digital Network Architecture Session Control Protocol (DNA SCP, used by IBM).

➤ *Transport*—Reliable end-to-end communication is the primary function of the transport layer. Its many responsibilities include:

  ➤ Ensuring flow control, so the amount of data being sent from one node will not overwhelm the destination node.

  ➤ Ensuring the ability of multiple connections to utilize a single transport (multiplexing).

  ➤ Ensuring the reliable transfer of data. The transport layer implements connection-oriented services between nodes, which utilize a three-way handshake (synchronization, acknowledgment, and data-transfer) to efficiently transfer data.

➤ Ensuring positive acknowledgment of the process of a node waiting for an acknowledgment from the destination node before sending data.

➤ Windowing, a form of flow control, that specifies how much data will be transferred between acknowledgments.

Transport layer protocols include Transmission Control Protocol (TCP, part of the TCP/IP protocol suite) and Sequenced Packet Exchange (SPX, part of the IPX/SPX protocol suite).

➤ *Network*—Selecting the appropriate path a packet should take to get to its intended destination is the function of the network layer. It is responsible for routing, which is the process of using a network layer address to determine the best path a packet will travel to its destination. Network layer protocols include Internet Protocol (IP, part of the TCP/IP protocol suite) and Internetwork Packet Exchange (IPX, part of the IPX/SPX protocol suite).

➤ *Data Link*—Divided into two sublayers (Media Access Control [MAC] and Logical Link Control [LLC]), the data-link layer is responsible for:

➤ Preparing data from upper layers to be transmitted over the physical medium by encapsulating upper-layer data into frames. This frame includes the source and destination of MAC addresses in the frame header.

➤ Converting data into bits, so it can be transmitted by the physical layer.

➤ Adding a cyclical redundancy check (CRC) to the end of a frame, which is used for error checking at the data-link layer.

Data-link protocols include High-Level Data Link Control (HDLC, the Cisco default encapsulation for serial connections), Synchronous Data Link Control, (SDLC, used in IBM networks), Link Access Procedure Balanced (LAPB, used with X.25), X.25 (packet switching network), Serial Line Internet Protocol (SLIP, an older TCP/IP dial-up protocol), Point-to-Point Protocol (PPP, a newer dial-up protocol), Integrated Services Digital Network (ISDN, a dial-up digital service), and frame relay (a packet switching network).

➤ *Physical*—The lowest layer of the OSI model, the physical layer defines the electrical functionality required to send and receive bits over a given physical medium. Specifications that define the voltage levels and physical components of a network are defined at the physical layer.

Protocols are not specified at the physical layer, because they are implemented as software. Examples of the standards for sending data over the physical medium are Ethernet (the most widely used standard in networking), Token Ring (IBM's proprietary networking topology), and Fiber Distributed Data Interface (FDDI, a standard for fiber optic networks, commonly used as a backbone).

For more information, see Chapter 4 of *CCNA Routing and Switching Exam Cram, Second Edition.*

Question 25

The correct answer is c. Flash is an erasable, programmable memory area that contains the Cisco operating system software (IOS). Answer a is incorrect, because ROM is a physical chip installed on a router's motherboard that contains the bootstrap program. Answer b is incorrect because NVRAM is used by Cisco routers to store the startup or running configuration. Answer d is incorrect, because RAM is used by a Cisco router as its main working area.

The internal components of Cisco routers are comparable to that of conventional PCs. A Cisco router is similar to a PC in that it contains an operating system (known as the IOS), a bootstrap program (like the BIOS used in a PC), RAM (memory), and a processor (CPU). Unlike a PC, a Cisco router has no external drives to install software or upgrade the IOS. Rather than use a floppy or CD drive to enter software as PCs do, Cisco routers are updated by downloading updated information from a TFTP server. The following list outlines the core components of Cisco routers:

➤ *Read-Only Memory (ROM)*—This is a physical chip installed on a router's motherboard that contains the bootstrap program, the POST (Power On Self Test), and the operating system software (Cisco IOS). When a Cisco router is first powered up, the bootstrap program and the POST are executed. ROM cannot be changed through software; the chip must be replaced if any modifications are needed.

➤ *Random Access Memory (RAM)*—As with PCs, RAM is used as the router's main working area and contains the running configuration. All information in RAM dissipates when the router is turned off.

➤ *Non-Volatile RAM (NVRAM)*—This component is used by Cisco routers to store the startup or running configuration. After a configuration is created, it exists and runs in RAM. Because RAM cannot permanently store this information, NVRAM is used. Information in NVRAM is preserved after the router is powered off.

➤ *Flash*—Essentially the programmable read-only memory (PROM) used in PCs, the Flash is an erasable, programmable memory area that contains the Cisco operating system software (IOS). When a router's operating system needs to be upgraded, new software can be downloaded from a TFTP server to the router's Flash. Upgrading the IOS in this manner is typically more convenient than replacing the ROM chip in the router's motherboard. Information in Flash is retained when the router is powered off.

For more information, see Chapter 8 of *CCNA Routing and Switching Exam Cram, Second Edition.*

Question 26

The correct answer is b. The logical circuit in a frame relay network is identified by the data-link connection identifier (DLCI). Answer a is an invalid choice, because ANSI is a standards organization, not an identifier in frame relay. Answers c and d are incorrect, because X.25 uses SVCs and PVCs, not frame relay. Packet switching is a wide area network (WAN) technology that sends data over WAN links through the use of logical circuits, where data is encapsulated into packets and routed by the service provider through various switching points. The two most common packet switching technologies in use are frame relay and X.25. The following list details the features of X.25 and frame relay:

➤ *Frame Relay*—This is a packet switching technology that utilizes logical circuits to form a connection. This logical connection is identified by the DLCI, which is a unique identifier used to identify a specific frame relay connection. Its connection is managed through a signaling format known as Local Management Interface (LMI). The LMI interface provides information about the DLCI values, as well as other connection-related information. Your Cisco router must utilize the same LMI signaling format as your service provider. LMI has three signaling formats: CISCO, ANSI, and Q933A. The Cisco default LMI type is CISCO.

➤ *X.25*—This is a packet switching technology that also utilizes logical or virtual circuits to facilitate WAN connectivity. The two types of logical circuits used in X.25 are switched virtual circuits (SVC) and permanent virtual circuits (PVC). An SVC is a temporary connection that is established on an as-needed basis and is closed when not in use. A PVC is a permanent connection that is always available for use.

For more information, see Chapter 11 of *CCNA Routing and Switching Exam Cram, Second Edition.*

Question 27

The correct answers are a, d, e, and g. Cisco routers support the use of access lists, which can be used to filter incoming or outgoing traffic. Each type of access list is assigned a range of numbers that distinguish one access list type from another. Answer b is incorrect, because a standard IP access list uses numbers between 1 and 99. Answer c is incorrect, because an extended IP access list uses numbers between 100 and 199. Answer f is incorrect, because a standard IPX access list uses numbers between 800 and 899. The following list details the correct numerical ranges used by common access lists:

➤ *Standard IP Access List*—1–99

➤ *Extended Access List*—100–199

➤ *Standard IPX Access List*—800–899

➤ *Extended IPX Access List*—900–999

➤ *IPX SAP Access List*—1000–1099

For more information, see Chapter 13 of *CCNA Routing and Switching Exam Cram, Second Edition.*

Question 28

The correct answer is a. A Cisco router's startup sequence is divided into three basic operations. The POST executes, the operating system is loaded, and, finally, the startup configuration is loaded into RAM. Answer b is incorrect, because there is no second POST. Answer c is incorrect, because the router's configuration is loaded into memory in the third step of the startup sequence, not the second. Answer d is incorrect, because the Flash does not perform a software check. The following details the startup sequence:

1. The bootstrap program in ROM executes the POST, which performs a basic hardware level check.

2. The Cisco IOS (operating systems software) is loaded into memory per instructions from the boot system command. The IOS can be loaded from Flash, ROM, or a network location (TFTP server).

3. The router's configuration file is loaded into memory from NVRAM. If no configuration file is found, the setup program automatically initiates the dialog necessary to create a configuration file.

For more information, see Chapter 8 of *CCNA Routing and Switching Exam Cram, Second Edition.*

Question 29

The correct answer is a. The OSI model defines standards used in networking and comprises a seven-layer model. The application layer of the OSI model functions as the "window" for applications to access network resources and is responsible for verifying the identity of the destination node. Since no other layer of the OSI model performs this function, all of the other answers are incorrect. The following list outlines the seven layers of the OSI model and their functions:

➤ *Application*—The "window" to networking used by programs, the application layer is responsible for:

  ➤ Verifying that the appropriate resources are present to initiate a connection with the destination node.

  ➤ Verifying the identity of the destination node.

Application layer protocols include Telnet, File Transfer Protocol (FTP), Simple Mail Transfer Protocol (SMTP), World Wide Web (WWW), Electronic Data Interchange (EDI), and Wide Area Information Server (WAIS).

➤ *Presentation*—Essentially a translator, the presentation layer is responsible for:

  ➤ Translating text and data syntax, such as Extended Binary-Coded Decimal Interchange Code (EBCDIC, used in IBM systems) to American Standard Code for Information Interchange (ASCII, used in PCs and most computer systems).

  ➤ Using Abstract Syntax Notation One (ASN 1) to perform data translation.

Presentation layer protocols include PICTure (PICT, Apple computer picture format), Tagged Image File Format (TIFF), Joint Photographic Experts Group (JPEG), Musical Instrument Digital Interface (MIDI), Motion Picture Expert Group (MPEG), and Quick Time (audio/video application).

➤ *Session*—Used to coordinate communication between nodes, the session layer is responsible for:

  ➤ Creating, maintaining, and ending a communication session.

  ➤ Coordinating service requests and responses that occur between nodes across the network.

Session layer protocols include Network File System (NFS, used by SUN Microsystems and Unix with TCP/IP), structured query language (SQL, used to define database information requests), remote procedure calls (RPCs, used in Microsoft network communication), X Window (used by Unix terminals), AppleTalk Session Protocol (ASP, used by Apple computers), and Digital Network Architecture Session Control Protocol (DNA SCP, used by IBM).

➤ *Transport*—Reliable end-to-end communication is the primary function of the transport layer. Its many responsibilities include:

➤ Ensuring flow control, so the amount of data being sent from one node will not overwhelm the destination node.

➤ Ensuring the ability of multiple connections to utilize a single transport (multiplexing).

➤ Ensuring the reliable transfer of data. The transport layer implements connection-oriented services between nodes, which utilize a three-way handshake (synchronization, acknowledgment, and data-transfer) to efficiently transfer data.

➤ Ensuring positive acknowledgment of the process of a node waiting for an acknowledgment from the destination node before sending data.

➤ Windowing, a form of flow control that specifies how much data will be transferred between acknowledgments.

Transport layer protocols include Transmission Control Protocol (TCP, part of the TCP/IP protocol suite) and Sequenced Packet Exchange (SPX, part of the IPX/SPX protocol suite).

➤ *Network*—Selecting the appropriate path a packet should take to get to its intended destination is the function of the network layer. It is responsible for routing, which is the process of using a network layer address to determine the best path a packet will travel to its destination. Network layer protocols include Internet Protocol (IP, part of the TCP/IP protocol suite) and Internetwork Packet Exchange (IPX, part of the IPX/SPX protocol suite).

➤ *Data Link*—Divided into two sublayers (Media Access Control [MAC] and Logical Link Control [LLC]), the data-link layer is responsible for:

➤ Preparing data from upper layers to be transmitted over the physical medium by encapsulating upper-layer data into frames. This frame includes the source and destination of MAC addresses in the frame header.

> ➤ Converting data into bits, so it can be transmitted by the physical layer.

> ➤ Adding a cyclical redundancy check (CRC) to the end of a frame, which is used for error checking at the data-link layer.

Data-link protocols include High-Level Data Link Control (HDLC, the Cisco default encapsulation for serial connections), Synchronous Data Link Control, (SDLC, used in IBM networks), Link Access Procedure Balanced (LAPB, used with X.25), X.25 (packet switching network), Serial Line Internet Protocol (SLIP, an older TCP/IP dial-up protocol), Point-to-Point Protocol (PPP, a newer dial-up protocol), Integrated Services Digital Network (ISDN, a dial-up digital service), and frame relay (a packet switching network).

> ➤ *Physical*—The lowest layer of the OSI model, the physical layer defines the electrical functionality required to send and receive bits over a given physical medium. Specifications that define the voltage levels and physical components of a network are defined at the physical layer. Protocols are not specified at the physical layer, because they are implemented as software. Examples of the standards for sending data over the physical medium are Ethernet (the most widely used standard in networking), Token Ring (IBM's proprietary networking topology), and Fiber Distributed Data Interface (FDDI, a standard for fiber optic networks, commonly used as a backbone).

For more information, see Chapter 4 of *CCNA Routing and Switching Exam Cram, Second Edition.*

Question 30

The correct answers are b, c, and d. Answer a is incorrect because a bridge resides at the data-link layer of the OSI model and can filter traffic based on the data-link address of a packet or MAC address. (The MAC address is the address burned into the network adapter card by the manufacturer.) Answer e is incorrect, because bridges pass unknown packets out all ports and also pass broadcast traffic. To prevent the potential loops that can occur from this broadcast traffic, bridges use the Spanning Tree Protocol, which allows for redundant connections between bridges while blocking traffic that can cause loops.

A router resides at the network layer of the OSI model and can filter traffic based on the network layer address (IP address). Routers do not pass broadcast traffic and will discard any packets with unknown destinations.

For more information, see Chapter 4 of *CCNA Routing and Switching Exam Cram, Second Edition.*

Question 31

The correct answer is d. **COPY RUNNING-CONFIG STARTUP-CONFIG** will copy the running configuration from RAM to the startup configuration file in NVRAM. The startup configuration is a configuration file in a Cisco router that contains the commands used to set router-specific parameters. The startup configuration is stored in NVRAM and is loaded into RAM when the router starts up. Answer a is incorrect, because the **COPY STARTUP-CONFIG RUNNING-CONFIG** command will copy the startup configuration to the running configuration, replacing the running configuration. Answer b is incorrect, because the running configuration is stored in RAM, not NVRAM. Answer c is incorrect, because the running configuration is stored in RAM, not Flash.

The running configuration is the startup configuration that is loaded into RAM and is the configuration used by the router when it is running. Because any information in RAM is lost when the router is powered off, the running configuration can be saved to the startup configuration, which is located in NVRAM, and will, therefore, be saved when the router is powered off. In addition, both the running and startup configuration files can be copied to and from a TFTP server, for backup purposes. The following list outlines the common commands used when working with configuration files:

➤ *COPY STARTUP-CONFIG RUNNING-CONFIG*—This command will copy the startup configuration file in NVRAM to the running configuration in RAM.

➤ *COPY STARTUP-CONFIG TFTP*—This command will copy the startup configuration file in NVRAM to a remote TFTP server.

➤ *COPY RUNNING-CONFIG STARTUP-CONFIG*—This command will copy the running configuration from RAM to the startup configuration file in NVRAM.

➤ *COPY RUNNING-CONFIG TFTP*—This command will copy the running configuration from RAM to a remote TFTP server.

➤ *COPY TFTP RUNNING-CONFIG*—This command will copy a configuration file from a TFTP server to the router's running configuration in RAM.

➤ *COPY TFTP STARTUP-CONFIG*—This command will copy a configuration file from a TFTP server to the startup configuration in NVRAM.

Note that the basic syntax of the **COPY** command specifies the source first, then the destination.

To modify a router's running configuration, you must be in global configuration mode. The **CONFIG T** command is used in privileged mode to access global configuration mode, which is displayed at the router prompt with "config" in parentheses **(RouterA(config)#)**. To exit global configuration mode, use the keys Control and Z (Ctrl+Z).

For more information, see Chapter 8 of *CCNA Routing and Switching Exam Cram, Second Edition.*

Question 32

The correct answers are a and d. Both RIP and IGRP are distance vector routing protocols. Answer b is incorrect, because Open Shortest Path First (OSPF) is a link state routing protocol. Answer c is incorrect, because Enhanced Interior Gateway Routing Protocol (EIGRP) is considered a balanced hybrid routing protocol that combines the features of link state and distance vector. Answers e and f are incorrect, because IP and IPX are routed protocols and are not used to update routing tables. A Cisco router cannot route anything unless an entry is in its routing table that contains information on where to send the packet.

A routing table can be of two basic types: static and dynamic. A static routing table is one that has its entries entered manually by an operator. On a Cisco router, a static route can be entered using the **IP ROUTE** command. A dynamic routing table is one that has its entries entered automatically by the router, through the use of a routing protocol. A dynamic routing protocol provides for automatic discovery of routes and eliminates the need for static routes. Routing protocols generally fall into one of two categories: link state and distance vector. Distance vector routing protocols dynamically update their routing tables by broadcasting their own routing information at specified intervals. Distance vector routing protocols include RIP and IGRP. Link state routing protocols, such as OSPF, update their routing tables by sending multicast "hello" messages to their neighbors and only send updated information when a change occurs in the network.

For more information, see Chapter 11 of *CCNA Routing and Switching Exam Cram, Second Edition.*

Question 33

The correct answer is d. All of these protocols are presentation layer protocols except answer d, Wide Area Information Server (WAIS). Therefore, answers a, b, c, e, f, and g are incorrect. WAIS is an application layer protocol. The application layer of the OSI model functions as the "window" for applications to access network resources and is responsible for verifying the identity of the destination node. Application layer protocols include Telnet, FTP, SMTP, WWW, EDI, and WAIS. The presentation layer functions as a translator and is responsible for text and data syntax translation, such as EBCDIC, used in IBM systems to ASCII, used in PC and most computer systems. Presentation layer protocols include PICT, TIFF, JPEG, MIDI, MPEG, and Quick Time (audio/video application).

For more information, see Chapter 4 of *CCNA Routing and Switching Exam Cram, Second Edition.*

Question 34

The correct answer is d. The access list **ACCESS-LIST 888 DENY 33 22** will deny IPX network 33 access to IPX network 22. Answer a is incorrect, because a standard IPX access list does not begin with "IPX". Answer b is incorrect, because the source and destination networks are in the wrong order. The syntax for a standard IPX access list specifies the source first, then the destination. Answer c is incorrect, because the numerical range for a standard IPX access list is 800–899.

Access lists can be used to filter the traffic that is handled by a Cisco router. An IPX access list will filter IPX traffic and can be created as a standard IPX access list or an extended IPX access list. A standard IPX access list can be used to permit or deny access based on the source and destination IPX address information. The syntax for creating a standard IPX access list is:

```
ACCESS-LIST [access-list-number] [permit|deny]
 source [source-node-mask] [destination]
 [destination-node-mask]
```

➤ *ACCESS-LIST*—This command is followed by the **access list number**. When using a standard IPX access list, this can be a number between 800 and 899.

➤ *[permit|deny]*—This statement will either allow or disallow traffic in this access list.

➤ *source*—This statement is used to specify the source IPX addressing information. A "-1" can be used to signify "all networks."

➤ *[destination]*—This statement is used to specify destination addressing information. (A "-1" may be used here as well.)

➤ *[source-node-mask] And [destination-node-mask]*—These are optional parameters for controlling access to a portion of a network, similar to the wildmask that is used in IP access lists.

For more information, see Chapter 13 of *CCNA Routing and Switching Exam Cram, Second Edition.*

Question 35

The correct answer is b. The **IPX ACCESS-GROUP 922 OUT** is the correct command to use in this question to apply the access list 922 to the outgoing interface of the router. Answer a is incorrect, because no hyphen is between **ACCESS** and **GROUP**. Answer c is incorrect, because the command must begin with IPX. Answer d is incorrect, because the correct command is **IPX ACCESS-GROUP**. For an IPX access list to filter traffic, it must be applied to an interface. This is done using the **IPX ACCESS-GROUP** command. This command must be entered in interface configuration mode. To access interface configuration mode, use the **INTERFACE [interface-number]** command from global configuration mode. When the router is in interface configuration mode, the router prompt displays "config-if" in parentheses, followed by a pound sign **((ROUTER(config-if)#)**. The **IPX ACCESS GROUP** command has the following syntax:

```
IPX ACCESS-GROUP [access-list number] [out|in]
```

The parameter **[out|in]** specifies where the access list will be applied on the router. **OUT** will cause the router to apply the access list to all outgoing packets, and **IN** will cause the router to apply the access list to all incoming packets.

Access lists can be used to filter the traffic that is handled by a Cisco router. An IP access list will filter IP traffic and can be created as a standard IP access list or an extended IP access list. A standard IP access list can be used to permit or deny access based on IP addressing information only and can only act on the source IP addressing information.

For more information, see Chapter 13 of *CCNA Routing and Switching Exam Cram, Second Edition.*

Question 36

The correct answer is b. User Datagram Protocol (UDP) is a transport layer protocol that is connectionless. Answer a is incorrect, because connectionless protocols do not create a virtual circuit. Answer c is incorrect, because connectionless protocols do not establish a communication session. Answer d is incorrect, because connectionless protocols do not utilize sequencing when communicating. Connectionless protocols feature:

➤ *No Sequencing or Virtual Circuit Creation*—Connectionless protocols do not sequence packets or create virtual circuits.

➤ *No Guarantee of Delivery*—Packets are sent as datagrams, and delivery is not guaranteed by connectionless protocols. When using a connectionless protocol like UDP, the guarantee of delivery is the responsibility of higher layer protocols.

➤ *Less Overhead*—Because connectionless protocols do not perform any of the above services, overhead is less than with connection-oriented protocols. TCP is a transport layer protocol that is connection-oriented. Connection-oriented protocols feature:

➤ Reliability that in a communication session is established between hosts before sending data. This session is considered a virtual circuit.

➤ Sequencing that occurs when a connection-oriented protocol like TCP sends data. It numbers each segment so that the destination host can receive the data in the proper order.

➤ Guaranteed delivery in connection-oriented protocols utilize error checking to guarantee packet delivery.

➤ More overhead because of the additional error checking and sequencing responsibilities. Connection-oriented protocols require more overhead than do connectionless protocols.

For more information, see Chapter 9 of *CCNA Routing and Switching Exam Cram, Second Edition.*

Question 37

The correct answer is a. Multiplexing is the ability of multiple connections to utilize a single transport and is not used as part of flow control. All of the other answers are incorrect, because they all describe valid methods of flow control. Flow control is a method used by TCP at the transport layer that ensures the amount of data being sent from one node will not overwhelm the destination node. Three basic techniques are used in flow control:

➤ *Windowing*—This is the process of predetermining the amount of data that will be sent before an acknowledgment is expected.

> *Buffering*—This is a method of temporarily storing packets in memory buffers until they can be processed by the computer.

> *Source-Quench Messaging*—This is a message sent by a receiving computer when its memory buffers are nearing capacity. A source-quench message is a device's way of telling the sending computer to slow down the rate of transmission.

For more information, see Chapter 9 of *CCNA Routing and Switching Exam Cram, Second Edition.*

Question 38

The correct answers are a, c, e, and f. The internal components of Cisco routers are comparable to that of conventional PCs. A Cisco router is similar to a PC in that it contains an operating system (known as the IOS), a bootstrap program (like the BIOS used in a PC), RAM (memory), and a processor (CPU). Answers b, d, and g are incorrect, because Cisco routers do not use these devices. Unlike a PC, a Cisco router has no external drives to install software or upgrade the IOS. Rather than use a floppy or CD drive to enter software as PCs do, Cisco routers are updated by downloading update information from a TFTP server. The following list outlines the core components of Cisco routers:

> *Read-Only Memory (ROM)*—This is a physical chip installed on a router's motherboard that contains the bootstrap program, the POST, and the operating system software (Cisco IOS). When a Cisco router is first powered up, the bootstrap program and the POST are executed. ROM cannot be changed through software; the chip must be replaced if any modifications are needed.

> *Random Access Memory (RAM)*—As with PCs, RAM is used as the router's main working area and contains the running configuration. All information in RAM dissipates when the router is turned off.

> *Non-Volatile RAM (NVRAM)*—This is used by Cisco routers to store the startup or running configuration. After a configuration is created, it exists and runs in RAM. Because RAM cannot permanently store this information, NVRAM is used. Information in NVRAM is preserved after the router is powered off.

> *Flash*—Essentially the PROM used in PCs, the Flash is an erasable, programmable memory area that contains the Cisco operating system software (IOS). When a router's operating system needs to be upgraded, new software can be downloaded from a TFTP server to the router's Flash. Upgrading the IOS in this manner is typically more convenient

than replacing the ROM chip in the router's motherboard. Information in Flash is retained when the router is powered off.

For more information, see Chapter 4 of *CCNA Routing and Switching Exam Cram, Second Edition.*

Question 39

The correct answer is a. When you first log into a router, you are placed in user mode. All of the other answers are incorrect, because the only mode you enter upon first logging into a router is user mode. The Cisco IOS uses a command interpreter known as **EXEC**, which contains two primary modes: user mode and privileged mode. User mode is used to display basic router system information and to connect to remote devices. The commands in user mode are limited in capability. In user mode, the router prompt is followed by an angle bracket (**ROUTER>**). Privileged mode is used to modify and view the configuration of the router, as well as to set IOS parameters. Privileged mode also contains the **CONFIGURE** command, which is used to access other configuration modes, such as global configuration mode and interface mode. In privileged mode, the router prompt is followed by a pound sign (**ROUTER#**). To enter privileged mode, enter the **ENABLE** command from user mode.

For more information, see Chapter 4 of *CCNA Routing and Switching Exam Cram, Second Edition.*

Question 40

The correct answer is a. To enter privileged mode, enter the **ENABLE** command from user mode. When you first log into a router, you are placed in user mode. In this mode, the router prompt is followed by an angle bracket (**ROUTER>**). In privileged mode, the router prompt is followed by a pound sign (**ROUTER#**). Answer b is incorrect, because the **CONFIGURE** command is not available in user mode. Answer c is incorrect, because the **ENABLE** command is not entered in privileged mode. In this mode, the router prompt is followed by a pound sign (**ROUTER#**). Answer d is incorrect, because the **CONFIGURE** command is used to enter global configuration mode, not privileged mode. The Cisco IOS uses a command interpreter known as EXEC, which contains two primary modes: user mode and privileged mode. User mode is used to display basic router system information and to connect to remote devices. The commands in user mode are limited in capability. Privileged mode is used to modify and view the configuration of the router, as well as to set IOS parameters. To enter privileged mode, enter the **ENABLE** command from user mode.

For more information, see Chapter 4 of *CCNA Routing and Switching Exam Cram, Second Edition.*

Question 41

The correct answer is b. The **CONFIGURE NET** command can be abbreviated as **CONFIG NET** and must be entered from privileged mode. In this mode, the router prompt is followed by a pound sign (**ROUTER#**). Answer a is incorrect, because the **CONFIGURE NET** command cannot be entered from user mode. In this mode, the router prompt is followed by an angle bracket (**ROUTER>**). Answer c is incorrect, because the prompt is already in configuration mode. After entering configuration mode, the router prompt displays "config" in parentheses (**ROUTER(config)#**). Answer d is incorrect, because the prompt is displaying interface mode (**ROUTER(CONFIG-IF)#**).

For more information, see Chapter 8 of *CCNA Routing and Switching Exam Cram, Second Edition.*

Question 42

The correct answer is c. Cisco routers can be configured to display a message when users log in. This message is known as a logon banner and can be configured using the **BANNER MOTD** command. All of the other answers are not valid Cisco IOS commands, therefore, answers a, b, and d are incorrect. The **BANNER MOTD** command must be entered in global configuration mode, which can be accessed from privileged mode using the **CONFIG T** command. When a router is in global configuration mode, the prompt displays "config" in parentheses (**Router(config)#**). The correct syntax for setting a logon banner is as follows:

```
BANNER MOTD # (Note that the pound sign
 [#] is a
delimiting character of your choice.) [Message text] #
```

An example of a logon banner would be:

```
Router(config)#banner motd #
Enter TEXT message. End with the
character '#'.
This is the message text of my logon banner #
Router(config)#^Z
```

For more information, see Chapter 8 of *CCNA Routing and Switching Exam Cram, Second Edition.*

Question 43

The correct answer is c. The **CLEAR ACCESS-LIST COUNTERS 800** command will clear the previous matches from access list 800. Answer a is incorrect, because in the **CLEAR ACCESS-LIST COUNTERS** command, no hyphen is between **LIST** and **COUNTERS**, and the command does not begin with IPX. Answer b is incorrect, because **MATCHES** is not a valid Cisco IOS command. Answer d is incorrect, because in the **CLEAR ACCESS-LIST COUNTERS** command, a hyphen is between **ACCESS** and **LIST**. The **SHOW ACCESS-LIST** command will display all access lists in use on the router and will also show each line of the access list and return the number of times a packet matched that line. To clear the matches, use the **CLEAR ACCESS-LIST COUNTERS** command. The syntax for the **CLEAR ACCESS-LIST COUNTERS** command is: **CLEAR ACCESS-LIST COUNTERS [access list number]**.

For more information, see Chapter 13 of *CCNA Routing and Switching Exam Cram, Second Edition.*

Question 44

The correct answer is c. Abbreviated as CO, the central office is the telephone company's office that acts as the central communication point for the customer. Answer a is incorrect, because the CPE is the collection of communication devices that exists at the customer's location. Answer b is incorrect, because the local loop is the cabling that extends from the DEMARC to the telephone company's location. Answer d is incorrect, because the DEMARC is the point at the customer's premises where CPE devices connect. The basic process of WAN communication begins with a call placed using Customer Premises Equipment (CPE). The CPE connects to the DEMARC, which passes the data through the local loop to the central office. The central office will act as the central distribution point for sending and receiving information. The following list details the common components of WAN communication:

➤ *Customer Premises Equipment (CPE)*—This is the collection of communication devices (telephones, modems, and so forth) that exists at the customer's location.

➤ *Demarcation (DEMARC)*—This is the point at the customer's premises where CPE devices connect. The telephone company owns the DEMARC, which is usually a large punch-down board located in a wiring closet.

➤ *Local Loop*—This is the cabling that extends from the DEMARC to the telephone company's CO.

➤ *Central Office (CO)*—This is the telephone company's office that acts as the central communication point for the customer.

For more information, see Chapter 11 of *CCNA Routing and Switching Exam Cram, Second Edition.*

Question 45

The correct answer is d. A switch is a data-link layer connectivity device used to segment a network. A switch resides at the data-link layer of the OSI model and can filter traffic based on the data-link address of a packet or MAC address. (The MAC address is the address burned into the network adapter card by the manufacturer.) Answers a and b are incorrect, because they are not valid switching methods. Answer c is incorrect because cut-through switching does not copy the packet to its buffer prior to sending it. Cisco switches employ two basic methods of forwarding packets: cut-through and store-and-forward. Store-and-forward switches will copy an incoming packet to the local buffer and perform error checking on the packet before sending it on to its destination. If the packet contains an error, it is discarded. Cut-through switches will read the destination address of an incoming packet and immediately search its switch table for a destination port. Cut-through switches will not perform any error checking of the packet. The fact that cut-through switches do not perform error checking or copy the packet to their buffer before processing means they experience less latency (delay) than store-and-forward switches.

For more information, see Chapter 13 of *CCNA Routing and Switching Exam Cram, Second Edition.*

Question 46

The correct answer is g. The OSI model defines standards used in networking and comprises a seven-layer model. The physical layer of the OSI model defines the electrical functionality required to send and receive bits over a given physical medium. Since no other layer of the OSI model performs this function, all of the other answers are incorrect. The following list outlines the seven layers of the OSI model and their functions:

➤ *Application*—The "window" to networking used by programs, the application layer is responsible for:

　➤ Verifying that the appropriate resources are present to initiate a connection with the destination node.

　➤ Verifying the identity of the destination node.

Application layer protocols include Telnet, File Transfer Protocol (FTP), Simple Mail Transfer Protocol (SMTP), World Wide Web (WWW), Electronic Data Interchange (EDI), and Wide Area Information Server (WAIS).

➤ *Presentation*—Essentially a translator, the presentation layer is responsible for:

> ➤ Translating text and data syntax, such as Extended Binary-Coded Decimal Interchange Code (EBCDIC, used in IBM systems) to American Standard Code for Information Interchange (ASCII, used in PCs and most computer systems).

> ➤ Using Abstract Syntax Notation One (ASN 1) to perform data translation.

Presentation layer protocols include PICTure (PICT, Apple computer picture format), Tagged Image File Format (TIFF), Joint Photographic Experts Group (JPEG), Musical Instrument Digital Interface (MIDI), Motion Picture Expert Group (MPEG), and Quick Time (audio/video application).

➤ *Session*—Used to coordinate communication between nodes, the session layer is responsible for:

> ➤ Creating, maintaining, and ending a communication session.

> ➤ Coordinating service requests and responses that occur between nodes across the network.

Session layer protocols include Network File System (NFS, used by SUN Microsystems and Unix with TCP/IP), structured query language (SQL, used to define database information requests), remote procedure calls (RPCs, used in Microsoft network communication), X Window (used by Unix terminals), AppleTalk Session Protocol (ASP, used by Apple computers), and Digital Network Architecture Session Control Protocol (DNA SCP, used by IBM).

➤ *Transport*—Reliable end-to-end communication is the primary function of the transport layer. Its many responsibilities include:

> ➤ Ensuring flow control, so the amount of data being sent from one node will not overwhelm the destination node.

> ➤ Ensuring the ability of multiple connections to utilize a single transport (multiplexing).

> ➤ Ensuring the reliable transfer of data. The transport layer implements connection-oriented services between nodes, which utilize a three-way handshake (synchronization, acknowledgment, and data-transfer) to efficiently transfer data.

➤ Ensuring positive acknowledgment of the process of a node waiting for an acknowledgment from the destination node before sending data.

➤ Windowing, a form of flow control that specifies how much data will be transferred between acknowledgments.

Transport layer protocols include Transmission Control Protocol (TCP, part of the TCP/IP protocol suite) and Sequenced Packet Exchange (SPX, part of the IPX/SPX protocol suite).

➤ *Network*—Selecting the appropriate path a packet should take to get to its intended destination is the function of the network layer. It is responsible for routing, which is the process of using a network layer address to determine the best path a packet will travel to its destination. Network layer protocols include Internet Protocol (IP, part of the TCP/IP protocol suite) and Internetwork Packet Exchange (IPX, part of the IPX/SPX protocol suite).

➤ *Data Link*—Divided into two sublayers (Media Access Control [MAC] and Logical Link Control [LLC]), the data-link layer is responsible for:

➤ Preparing data from upper layers to be transmitted over the physical medium by encapsulating upper-layer data into frames. This frame includes the source and destination of MAC addresses in the frame header.

➤ Converting data into bits, so it can be transmitted by the physical layer.

➤ Adding a cyclical redundancy check (CRC) to the end of a frame, which is used for error checking at the data-link layer.

Data-link protocols include High-Level Data Link Control (HDLC, the Cisco default encapsulation for serial connections), Synchronous Data Link Control, (SDLC, used in IBM networks), Link Access Procedure Balanced (LAPB, used with X.25), X.25 (packet switching network), Serial Line Internet Protocol (SLIP, an older TCP/IP dial-up protocol), Point-to-Point Protocol (PPP, a newer dial-up protocol), Integrated Services Digital Network (ISDN, a dial-up digital service), and frame relay (a packet switching network).

➤ *Physical*—The lowest layer of the OSI model, the physical layer defines the electrical functionality required to send and receive bits over a given physical medium. Specifications that define the voltage levels and physical components of a network are defined at the physical layer. Protocols are not specified at the physical layer, because they are implemented as software. Examples of the standards for sending data over the physical medium are Ethernet (the most widely used standard in networking), Token Ring (IBM's proprietary networking topology), and Fiber Distributed Data Interface (FDDI, a standard for fiber optic networks, commonly used as a backbone).

For more information, see Chapter 4 of *CCNA Routing and Switching Exam Cram, Second Edition.*

Question 47

The correct answers are a, b, and d. A data-link address is also known as a physical address, hardware address, or more commonly, a MAC address. (So named a MAC address because this address resides at the MAC, or Media Access Control, sublayer of the data-link layer.) The MAC address is the address "burned" into every network adapter card by the manufacturer. This address is considered a "flat" address, because no logical arrangement of these addresses is on a network. A flat addressing scheme simply gives each member a unique identifier that is associated with him or her.

A network address, or logical address, is an address that resides at the network layer of the OSI model. Network addresses are hierarchical in nature, and a hierarchical addressing scheme uses logically structured addresses to provide a more organized environment. Examples of hierarchical network addresses are IP and IPX. An IP address comprises a network ID, which identifies the network a host belongs to, and a host ID, which is unique to that host. An IPX address utilizes a similar network ID and host ID format in addressing. The hierarchical addressing of IP and IPX enable complex networks to be logically grouped and organized.

For more information, see Chapter 9 of *CCNA Routing and Switching Exam Cram, Second Edition.*

Question 48

The correct answer is a. Based on the display, the next command to issue to set the virtual terminal password to **EXAM** would be **PASSWORD EXAM**. All of the other answers are not valid Cisco IOS commands, therefore, answers b, c, and d are incorrect. Cisco routers utilize passwords to provide security access to a router. Passwords can be set for controlling access to privileged mode, for remote sessions via Telnet, or for access through the auxiliary port. The virtual terminal password is used to control remote Telnet access to a router. Setting the virtual terminal password will require all users that Telnet into the router to provide this password for access. To configure the virtual terminal password, use the following commands:

```
LINE VTY 0 4
LOGIN
PASSWORD [password]
```

For more information, see Chapter 8 of *CCNA Routing and Switching Exam Cram, Second Edition.*

Question 49

The correct answer is b. Interface configuration mode is used to set interface-specific parameters on a Cisco router. To access interface configuration mode, the **INTERFACE** command is used. When issuing the **INTERFACE** command, it must be followed by the interface ID that is to be accessed (**S0** for serial port 0, **S1** for serial port 1, **E0** for Ethernet port 0, and so forth). The **INTERFACE** command must be issued from global configuration mode. When a router is in global configuration mode, the prompt displays "config" in parentheses (**Router(config)#**). All of the other answers, a, c, and d, are incorrect, because the router is not in global configuration mode. The **SHOW INTERFACE** command is used to display the configuration information of all interfaces. If the **SHOW INTERFACE** is issued with a specific port (that is, **SHOW INTERFACE E0**), then only the configuration of that interface will be displayed. The **SHOW INTERFACE** command can be issued from user or privileged mode.

For more information, see Chapter 8 of *CCNA Routing and Switching Exam Cram, Second Edition.*

Question 50

The correct answer is b. Answer b is the only valid standard IP access list. Answer a is incorrect, because the access list is using number 100, which is used in extended IP access lists. Answer c is incorrect, because no hyphen is between **ACCESS** and **LIST**. Answer d is incorrect, because no access list number is specified.

Access lists can be used to filter the traffic that is handled by a Cisco router. An IP access list will filter IP traffic and can be created as either a standard IP access list or an extended IP access list. A standard IP access list can be used to permit or deny access based on IP addressing information only, and it can only act on the source IP addressing information. The syntax for creating a standard IP access list is:

```
ACCESS-LIST [access-list-number] [permit|deny]
 source
 [wildcard mask]
```

➤ *ACCESS-LIST*—This command is followed by the **access list number.** When using a standard IP access list, this can be a number between 1 and 99.

➤ *[permit|deny]*—This statement will either allow or disallow traffic in this access list.

➤ *source*—This statement is used to specify the source IP addressing information. A standard IP access list can only act on a source address.

➤ *[wildcard mask]*—This is an optional parameter in a standard IP access list. If no **wildcard mask** is specified, the access list will only act on a single IP address that is specified in the **source.** A wildmask can be used to control access to an entire IP network ID rather than a single IP address; it will use the number 255 to mean "any" and the number 0 to mean "must match exactly."

For more information, see Chapter 13 of *CCNA Routing and Switching Exam Cram, Second Edition.*

Question 51

The correct answer is a. The access list **ACCESS-LIST 100 PERMIT TCP 200.200.200.50 0.0.0.0 131.200.0.0 0.0 255.255 EQ 23** will allow host **200.200.200.50** to have Telnet access to network **131.200.0.0.** Answer b is incorrect, because the access list number is 10, which is used for standard access lists, not extended; extended access lists use numbers between 100 and 199. Answer c is incorrect, because the port number used is 21, which specifies FTP traffic, not Telnet traffic. Telnet uses port 23. Answer d is incorrect, because no hyphen is between **ACCESS** and **LIST.**

The syntax for creating an extended IP access list is as follows:

```
ACCESS-LIST [access-list-number] [permit|deny]
 [protocol] [source]
 [wildcard mask] [destination] [wildmask]
 [operator][port]
```

➤ *ACCESS-LIST*—This command denotes an access list.

➤ *[access-list-number]*—This can be a number between 100 and 199 that is used to denote an extended IP access list.

➤ *[permit\deny]*—This will allow (permit) or disallow (deny) traffic specified in this access list.

➤ *[protocol]*—This parameter is used to specify the protocol that will be filtered in this access list. Common values are TCP, UDP, and IP (use of IP will denote all IP protocols).

➤ *[source]*—This parameter specifies the source IP addressing information.

➤ *[wildcard mask]*—This parameter can be optionally applied to further define the **source**. A wildmask can be used to control access to an entire IP network ID rather than a single IP address. A **wildcard mask** will use the number 255 to mean "any" and the number 0 to mean "must match exactly." The terms *host* and *any* may also be used here to more quickly specify a single host or an entire network. The **HOST** keyword is the same as the wildcard mask 0.0.0.0; the **ANY** keyword is the same as the wildcard mask 255.255.255.255.

➤ *[destination]*—This specifies the destination IP addressing information.

➤ *[wildmask]*—This parameter can be optionally applied to further define the destination IP addressing.

➤ *[operator]*—This statement can be optionally applied to define how to interpret the value entered in the **[port]** section of the access list. Common values are **EQ** (equal to the port specified), **LT** (less than the port number specified), and **GT** (greater than the port number specified).

➤ *[port]*—This specifies the port number this access list will act on. Common port numbers include 21 (FTP), 23 (TELNET), 25 (SMTP), 53 (DNS), 69 (TFTP), and 80 (HTTP).

For more information, see Chapter 13 of *CCNA Routing and Switching Exam Cram, Second Edition*.

Question 52

The correct answers are b, c, and d. Answer a is incorrect, because the Ethernet II frame type is called ARPA in Cisco. Answer e is incorrect, because the Token Ring frame type is called Token in Cisco. Answer f is incorrect, because the Token_Ring_Snap frame type is called Snap in Cisco. The IPX/SPX protocol is used in Novell NetWare networks. When IPX packets are passed to the data-link layer, they are encapsulated into frames for transmission over the physical media. Novell NetWare supports a number of different encapsulation methods, or frame types, used to encapsulate IPX packets. The frame type used by NetWare computers must be the same as the one used by a Cisco router, or they will not be able to communicate. When configuring a Cisco router to support IPX, you can specify the frame type to use. Novell NetWare and Cisco both use different names to describe the same frame type. The following list details the frame types used by NetWare networks and the comparable Cisco name.

➤ *ETHERNET II*—This is used by Novell and is called ARPA in Cisco.

➤ *ETHERNET 802.2*—This is used by Novell and is called SAP in Cisco.

➤ *ETHERNET_snap*—This is used by Novell and is called SNAP in Cisco.

➤ *ETHERNET 802*—This is used by Novell and is called NOVELL-ETHER in Cisco and is the default frame type.

➤ *TOKEN RING*—This is used by Novell and is called TOKEN in Cisco.

➤ *TOKEN RING_SNAP*—This is used by Novell and is called SNAP in Cisco.

For more information, see Chapter 12 of *CCNA Routing and Switching Exam Cram, Second Edition.*

Question 53

The correct answers are a and b. The Cisco IOS includes a command history feature that can be used to store previously entered commands in a history buffer. Answer c is incorrect, because the command history is enabled by default, not disabled. Answer d is incorrect, because the command history holds 10 entries by default, and can be configured to hold up to 256. Commands in the history buffer can be recalled by an operator at the router prompt, using Ctrl+P and Ctrl+N, saving configuration time. The command history feature is enabled by default, but it can be disabled, and the size of the history buffer can be modified. The following list details the common commands used when modifying the command history:

➤ *SHOW HISTORY*—Typing the **SHOW HISTORY** command will display all of the commands currently stored in the history buffer. Pressing Ctrl+P will recall previous commands in the command history at the router prompt. (The up arrow will do this as well.) Pressing

Ctrl+N will return more recent commands in the command history after recalling previous commands. (The down arrow will do this as well.)

➤ *TERMINAL HISTORY SIZE (0-256)*—Typing "**TERMINAL HISTORY SIZE (0-256)**" will allow you to set the number of command lines the command history buffer will hold. You must enter a number between 0 and 256. (The default is 10 commands.)

➤ *TERMINAL NO EDITING*—Typing "**TERMINAL NO EDITING**" will disable the command history feature. It is enabled by default.

➤ *TERMINAL EDITING*—Typing "**TERMINAL EDITING**" will enable the command history feature if it has been disabled.

For more information, see Chapter 8 of *CCNA Routing and Switching Exam Cram, Second Edition.*

Question 54

The correct answer is b. **COPY STARTUP-CONFIG TFTP** will copy the startup configuration file in NVRAM to a remote TFTP server. Answer a is incorrect, because the syntax for the **COPY STARTUP-CONFIG TFTP** command has a hyphen between **STARTUP** and **CONFIG**. Answer c is incorrect, because it is not a valid Cisco command. Answer d is incorrect, because **COPY TFTP STARTUP-CONFIG** will copy a configuration file from a TFTP server to the startup configuration in NVRAM.

The startup configuration is a configuration file in a Cisco router that contains the commands used to set router-specific parameters. The startup configuration is stored in NVRAM and is loaded into RAM when the router starts up.

The running configuration is the startup configuration that is loaded into RAM and is the configuration used by the router when it is running. Because any information in RAM is lost when the router is powered off, the running configuration can be saved to the startup configuration, which is located in NVRAM, and will, therefore, be saved when the router is powered off. In addition, both the running and startup configuration files can be copied to a TFTP server for backup purposes. The following list outlines the common commands used when working with configuration files:

➤ *COPY STARTUP-CONFIG RUNNING-CONFIG*—This command will copy the startup configuration file in NVRAM to the running configuration in RAM.

➤ *COPY STARTUP-CONFIG TFTP*—This command will copy the startup configuration file in NVRAM to a remote TFTP server.

➤ *COPY RUNNING-CONFIG STARTUP-CONFIG*—This command will copy the running configuration from RAM to the startup configuration file in NVRAM.

➤ *COPY RUNNING-CONFIG TFTP*—This command will copy the running configuration from RAM to a remote TFTP server.

➤ *COPY TFTP RUNNING-CONFIG*—This command will copy a configuration file from a TFTP server to the router's running configuration in RAM.

➤ *COPY TFTP STARTUP-CONFIG*—This command will copy a configuration file from a TFTP server to the startup configuration in NVRAM.

Note that the basic syntax of the **COPY** command specifies the source first, then the destination.

To modify a router's running configuration, you must be in global configuration mode. The **CONFIG T** command is used in privileged mode to access global configuration mode. Global configuration mode is displayed at the router prompt with "config" in parentheses **(RouterA(config)#)**. To exit global configuration mode, use the keys Control and Z (Ctrl+Z).

For more information, see Chapter 8 of *CCNA Routing and Switching Exam Cram, Second Edition.*

Question 55

The correct answer is b. Answer a is incorrect, because the subnet mask is 255.255.255.224. Answer c is incorrect, because this is a Class B address, not a Class c address. Answer d is incorrect, because the subnet mask is 255.255.255.224. When configuring IP on a Cisco router interface, you must specify the IP address for the interface and the number of bits used in the subnet mask. When an IP address is specified for the interface, the router will identify the address by the first octet rule as either a Class A, B, or C address. (Class A=1–126, Class B=28–191, Class C=192–223). The default subnet mask for the address class is assumed to be preset by the router. (Class A=255.0.0.0, Class B=255.255.0.0, Class C=255.255.255.0.) The second step to configuring IP on a router interface involves specifying the number of bits in the subnet field. If this number is 0, the default subnet mask is applied based on the address class. The number of bits specified in the subnet field will be applied by the router to the host portion of the IP address. In this question, a Class B address was entered for the interface (130.200.200.254). The router then assumed a subnet mask of 255.255.0.0, the default subnet mask for a Class B address. The number of bits in the subnet field was specified as 11, which was interpreted by the router as 8 bits in the *third* octet of the IP address and 3 bits in the *fourth* octet of the IP address and is applied as 255.255.255.224. (Eight bits in binary=11111111, which is the decimal equivalent to 255, and 3 bits in binary=11100000, which is the decimal equivalent to 224.)

For more information, see Chapter 8 of *CCNA Routing and Switching Exam Cram, Second Edition.*

Question 56

The correct answer is b. **CONFIGURE OVERWRITE-NETWORK (CONFIG O)** is used to copy a configuration file into NVRAM from a TFTP server. Use of the **CONFIGURE OVERWRITE-NETWORK** will not alter the running configuration. Answer a is incorrect, because **CONFIGURE TERMINAL** is used to enter configuration commands into the router from the console port or through Telnet. Answer c is incorrect, because **CONFIGURE OVERWRITE NETWORK** is used to copy a configuration file into NVRAM from a TFTP server. Answer d is incorrect, because **CONFIGURE MEMORY** is used to execute the configuration stored in NVRAM.

The **CONFIGURE** command, executed in privileged mode, is used to enter the configuration mode of a Cisco router. The four parameters used with the **CONFIGURE** command are **TERMINAL, MEMORY, OVERWRITE-NETWORK**, and **NETWORK**. The following list outlines the functions of these parameters:

➤ *CONFIGURE TERMINAL (CONFIG T)*—This command is used to enter configuration commands into the router from the console port or through Telnet.

➤ *CONFIGURE NETWORK (CONFIG NET)*—This command is used to copy the configuration file from a TFTP server into the router's RAM.

➤ *CONFIGURE OVERWRITE-NETWORK (CONFIG O)*—This command is used to copy a configuration file into NVRAM from a TFTP server. Use of the **CONFIGURE OVERWRITE-NETWORK** will not alter the running configuration.

➤ *CONFIGURE MEMORY (CONFIG MEM)*—This command is used to execute the configuration stored in NVRAM. It will copy the startup configuration to the running configuration.

For more information, see Chapter 8 of *CCNA Routing and Switching Exam Cram, Second Edition.*

Question 57

The correct answers are b and g. The only protocols in this question that reside at the transport layer are SPX and TCP. All of the other answers are incorrect, because they are not transport layer protocols. Transport layer protocols include part of the TCP/IP protocol suite and SPX, part of the IPX/SPX protocol suite. Session layer protocols include NFS, used by SUN Microsystems and Unix with TCP/IP; SQL, used to define database information requests; RPC, used in Microsoft network communication; X Window, used by Unix terminals; ASP, used by Apple computers; and DNA SCP, used by IBM. SMTP and Telnet are application layer protocols. PICT and TIFF are presentation layer protocols.

For more information, see Chapter 4 of *CCNA Routing and Switching Exam Cram, Second Edition*.

Question 58

The correct answers are b and e. The **SHOW RUNNING-CONFIG** command can be used in privileged mode to display the current running configuration in RAM and can be abbreviated as **SH RUN**. Answers a, c, and f are incorrect, because the command is entered in user mode. In this mode, the router prompt is followed by an angle bracket (**ROUTER>**). The **SHOW RUNNING-CONFIG** command cannot be used in either user or global configuration modes. In privileged mode, the router prompt is followed by a pound sign (**ROUTER#**). Answer d is incorrect, because the correct syntax for the **SHOW RUNNING-CONFIG** command uses a hyphen between **RUNNING** and **CONFIG**. Answer g is incorrect, because **DISPLAY RUNNING** is not a valid Cisco IOS command.

For more information, see Chapter 8 of *CCNA Routing and Switching Exam Cram, Second Edition*.

Question 59

The correct answer is d. The enable password is used to control access to the privileged mode of the router. Cisco routers utilize passwords to provide security access to a router. Passwords can be set for controlling access to privileged mode, for remote sessions via Telnet, or for access through the auxiliary port. The following list details the common passwords used in Cisco routers:

➤ *Enable Password*—The enable password is used to control access to the privileged mode of the router. If an enable password is specified, it must be entered after the **ENABLE** command to successfully access privileged mode. To configure the enable password, use the **ENABLE PASS-WORD** [*password*] command.

➤ *Enable Secret Password*—The enable secret password is used to control access to privileged mode, similar to the enable password. The difference between enable secret and enable password is the enable secret password will be encrypted for additional security. The enable secret password will take precedence over the enable password if both are enabled. To configure the enable secret password, use the **ENABLE SECRET [password]** command.

➤ *Virtual Terminal Password*—The virtual terminal password is used to control remote Telnet access to a router. Setting the virtual terminal password will require all users that Telnet into the router to provide this password for access. To configure the virtual terminal password, use the following commands:

```
LINE VTY 0 4
LOGIN
PASSWORD [password]
```

➤ *Auxiliary Password*—The auxiliary password is used to control access to any auxiliary ports the router may have. Setting the auxiliary password will require all users connecting to the auxiliary port of the router (usually remote dial-in) to provide this password for access. To configure the auxiliary password, use the following commands:

```
LINE AUX 0
LOGIN
PASSWORD [password]
```

➤ *Console Password*—The console password is used to control access to the console port of the router. Setting the console password will require all users connecting to the router console to provide this password for access. To configure the console password, use the following commands:

```
LINE CON 0
LOGIN
PASSWORD [password]
```

For more information, see Chapter 8 of *CCNA Routing and Switching Exam Cram, Second Edition*.

Question 60

The correct answer is c. Answers a, b, and d are incorrect, because HDLC is a data-link only layer protocol. Two common protocols used when encapsulating data over serial links are SDLC and HDLC. HDLC is an ISO standard data-link protocol used in WAN communication. Cisco routers use HDLC as the default protocol for all serial (WAN) links. SDLC is a data-link layer protocol used by IBM networks when communicating over WAN links using the Systems Network Architecture (SNA) protocol. Answers a, b, and d are incorrect, because HDLC is a data-link layer protocol.

For more information, see Chapter 4 of *CCNA Routing and Switching Exam Cram, Second Edition.*

Question 61

The correct answers are a, c, and e. IGRP is a distance vector protocol that is proprietary to Cisco and was designed to address the shortcomings of RIP. IGRP has a maximum hop count of 255 and can make more informed decisions when selecting routes. Answer b is incorrect, because IGRP considers hop count, delay, bandwidth, and reliability when determining the best path, making it a more efficient routing protocol than RIP. Answer d is incorrect, because the maximum hop count used by IGRP is 255. Answer f is incorrect, because IGRP is a distance vector protocol.

RIP is a dynamic routing protocol used to automatically update entries in a routing table and is a distance vector routing protocol. RIP routers will broadcast the contents of their routing table every 30 seconds. This broadcast traffic does not make RIP very efficient for large networks. To select the best route, a RIP router selects the path with the fewest number of hops. (One hop is the distance from one router to another.) The maximum hop count used in RIP is 15. Any destination that requires more than 15 hops is considered unreachable by RIP.

For more information, see Chapter 11 of *CCNA Routing and Switching Exam Cram, Second Edition.*

Question 62

The correct answer is b. The IPX SAP access list **ACCESS-LIST 1000 DENY –1 7** will prevent a Cisco router from passing print service advertisements sent from all NetWare servers. Answer a is incorrect, because no hyphen is between **ACCESS** and **LIST**. Answers c and d are incorrect, because the service type is 4, which specifies file services, not print services. An IPX SAP filter can be applied to a router's interface to filter the Service Advertising Protocol (SAP) advertisement traffic passed by the router. Novell NetWare servers will broadcast a list of their available resources every 60 seconds. This is known as a SAP advertisement. Cisco routers will record this information into their own SAP table for advertisement to other segments. By filtering the amount of SAP traffic with an IPX SAP access list, network traffic can be controlled and/or reduced. The syntax for a SAP access list is as follows:

```
ACCESS-LIST [access-list-number] [permit|deny]
  [source network]
  [service type]
```

➤ *ACCESS-LIST*—This command is followed by the **access list number**. When using an IPX SAP access list, this can be a number between 1000 and 1099.

➤ *[permit|deny]*—This statement will either allow or disallow traffic in this access list.

➤ *[source network]*—This statement is used to specify the source IPX network. A "-1" can be used to signify "all networks."

➤ *[service type]*—This statement is used to specify the type of service to be filtered. Common IPX service types include a NetWare file server, specified as number 4, and a NetWare printer server, specified as number 7.

For more information, see Chapter 13 of *CCNA Routing and Switching Exam Cram, Second Edition.*

Question 63

The correct answers are b and d. X.25 is a packet switching technology that also utilizes logical, or virtual, circuits to facilitate WAN connectivity. The two types of logical circuits used in X.25 are SVC and PVC. Answer a is incorrect, because an SVC is a temporary connection that is established on an as-needed basis and is closed when not in use. Answer c is incorrect, because a PVC is a permanent connection that is always available for use.

For more information, see Chapter 11 of *CCNA Routing and Switching Exam Cram, Second Edition.*

Question 64

The correct answer is b. ISDN is a communications standard that uses digital telephone lines to transmit voice, data, and video. It is a dial-up service that is used on demand. ISDN protocol standards are defined by the ITU. Protocols that begin with the letter E (such as E.164) define ISDN standards on the existing telephone network. Protocols that begin with the letter Q (such as Q.931 and Q.921) specify ISDN switching and signaling standards, and protocols that begin with the letter I (such as I.430 and I.431) specify ISDN concepts, interfaces, and terminology. Answer a is incorrect, because protocols that begin with the letter Q (such as Q.931 and Q.921) specify ISDN switching and signaling standards. Answer c is incorrect, because protocols that begin with the letter I (such as I.430 and I.431) specify ISDN concepts, interfaces, and terminology. Answer d is incorrect, because protocols that begin with the letter E (such as E.164) define ISDN standards on the existing telephone network.

For more information, see Chapter 11 of *CCNA Routing and Switching Exam Cram, Second Edition*.

Question 65

The correct answer is d. The IP address range 129.23.98.25–129.23.98.30 is a valid host ID for the nodes in this subnet. Answers a, b, and c are incorrect, because the correct address range begins with 129.23.98.24. When subnetting IP addresses, host IDs are grouped together according to the incremental value of the subnet mask in use. The first address in this range identifies the subnet (subnet ID) and is not used for a host ID. The last address in this range is used by the router as the broadcast address, also known as the default gateway. In this question, the subnet mask used is 248, which will divide subnets in groups of 8. (129.23.98.8–129.23.98.15, 129.23.98.16–129.23.98.23, 129.23.98.24–129.23.98.31, and so forth.) The broadcast address is specified as 129.23.98.31, which places this address in subnet 129.23.98.24–129.23.98.31. The first address is used as the subnet ID and the last address is used for the broadcast address, making the valid range of host IDs 129.23.98.25–129.23.98.30. The following list details the incremental values of IP subnet masks:

➤ *Subnet Mask 192*—Increments in 64

➤ *Subnet Mask 224*—Increments in 32

➤ *Subnet Mask 240*—Increments in 16

➤ *Subnet Mask 248*—Increments in 8

➤ *Subnet Mask 252*—Increments in 4

➤ *Subnet Mask 254*—Increments in 2

➤ *Subnet Mask 255*—Increments in 1

For more information, see Chapter 8 of *CCNA Routing and Switching Exam Cram, Second Edition.*

Question 66

The correct answer is d. NVRAM is used by Cisco routers to store the startup or running configuration. Answer a is incorrect, because the Flash is an erasable, programmable memory area that contains the Cisco operating system software. Answer b is incorrect, because RAM is used as the routers' main working area, and does not store any information when the router is powered down. Answer c is incorrect, because ROM is a physical chip installed on a router's motherboard that contains the bootstrap program. The internal components of Cisco routers are comparable to that of conventional PCs. A Cisco router is similar to a PC in that it contains an operating system (known as the IOS), a bootstrap program (like the BIOS used in a PC), RAM (memory), and a processor (CPU). Unlike a PC, a Cisco router has no external drives to install software or upgrade the IOS. Rather than use a floppy or CD drive to enter software as PCs do, Cisco routers are updated by downloading update information from a TFTP server. The following list outlines the core components of Cisco routers:

➤ *Read-Only Memory (ROM)*—This is a physical chip installed on a router's motherboard that contains the bootstrap program, the POST, and the operating system software (Cisco IOS). When a Cisco router is first powered up, the bootstrap program and the POST are executed. ROM cannot be changed through software; the chip must be replaced if any modifications are needed.

➤ *RAM (Random Access Memory)*—As with PCs, RAM is used as the router's main working area and contains the running configuration. All information in RAM dissipates when the router is turned off.

➤ *Non-Volatile RAM (NVRAM)*—This is used by Cisco routers to store the startup or running configuration. After a configuration is created, it exists and runs in RAM. Because RAM cannot permanently store this information, NVRAM is used. Information in NVRAM is preserved after the router is powered off.

➤ *Flash*—This is essentially the PROM used in PCs. The Flash is an erasable, programmable memory area that contains the Cisco operating system software (IOS). When a router's operating system needs to be upgraded, new software can be downloaded from a TFTP server to the router's Flash. Upgrading the IOS in this manner is typically more convenient than replacing the ROM chip in the router's motherboard. Information in Flash is retained when the router is powered off.

For more information, see Chapter 4 of *CCNA Routing and Switching Exam Cram, Second Edition.*

Question 67

The correct answer is d. The router's configuration file is loaded into memory from NVRAM. If no configuration file is found, the setup program automatically initiates the dialog necessary to create a configuration file. Based on this information, all of the other answers are incorrect. ROM contains the IOS and the bootstrap program, not the configuration file. Flash contains the Cisco IOS, not the configuration file. Because the router will automatically initiate the setup program, it will not crash in the absence of a configuration file.

For more information, see Chapter 4 of *CCNA Routing and Switching Exam Cram, Second Edition.*

Question 68

The correct answers are a, c, d, and e. The **LOGOUT** command is used to log out of a router from privileged mode. The **QUIT, EXIT,** and **DISABLE** commands may also be used to log out of privileged mode. **LOGOFF, LEAVE-OFF,** and **QUIT-ROUTER [ID]** are not valid Cisco IOS commands, therefore, answers b, f, and g are incorrect. The Cisco IOS uses a command interpreter known as EXEC, which contains two primary modes: user mode and privileged mode. User mode is used to display basic router system information and to connect to remote devices. The commands in user mode are limited in capability. In user mode, the router prompt is followed by an angle bracket **(ROUTER>)**. Privileged mode is used to modify and view the configuration of the router, as well as to set IOS parameters. Privileged mode also contains the **CONFIGURE** command, which is used to access other configuration modes, such as global configuration mode and interface mode. In privileged mode, the router prompt is followed by a pound sign **(ROUTER#)**. To enter privileged mode, enter the **ENABLE** command from user mode.

For more information, see Chapter 8 of *CCNA Routing and Switching Exam Cram, Second Edition.*

Question 69

The correct answer is e. Pressing the Escape key along with the letter *B* will move the cursor back one word. None of the other answers will move the cursor back one word.

Configuring a Cisco router can sometimes involve long, detailed command lines that can be slow to navigate. For this reason, the Cisco IOS includes a number of shortcuts through them. The following list details some of the more common methods used to navigate the command line:

➤ *Ctrl+A*—Pressing the Control key along with the letter *A* will move the cursor to the beginning of the command line.

➤ *Ctrl+E*—Pressing the Control key along with the letter *E* will move the cursor to the end of the command line.

➤ *Ctrl+P*—Pressing the Control key along with the letter *P* will recall previous commands in the command history. (The up arrow will do this as well.)

➤ *Ctrl+N*—Pressing the Control key along with the letter *N* will return to more recent commands in the command history after recalling previous commands. (The down arrow will do this as well.)

➤ *Esc+B*—Pressing the Escape key along with the letter *B* will move the cursor back one word.

➤ *Esc+F*—Pressing the Escape key along with the letter *F* will move the cursor forward one word.

➤ *Left and Right Arrow Keys*—These arrow keys will move the cursor one character left and right, respectively.

➤ *Tab*—Pressing the Tab key will complete an entry typed at the router prompt.

For more information, see Chapter 8 of *CCNA Routing and Switching Exam Cram, Second Edition.*

Question 70

The correct answer is e. The console password is used to control access to the console port of the router. Cisco routers utilize passwords to provide security access to a router. Passwords can be set for controlling access to privileged mode, accessing remote sessions via Telnet, or controlling access through the auxiliary port. The following list details the common passwords used in Cisco routers:

➤ *Enable Password*—The enable password is used to control access to the privileged mode of the router. If an enable password is specified, it must be entered after the **ENABLE** command to successfully access privileged mode. To configure the enable password, use the **ENABLE PASS-WORD** [*password*] command.

➤ *Enable Secret Password*—The enable secret password is used to control access to privileged mode, similar to the enable password. The difference between enable secret and enable password is that the enable secret password will be encrypted for additional security. The enable secret password will take precedence over the enable password if both are enabled. To configure the enable secret password, use the **ENABLE SECRET [password]** command.

➤ *Virtual Terminal Password*—The virtual terminal password is used to control remote Telnet access to a router. Setting the virtual terminal password will require all users that Telnet into the router to provide this password for access. To configure the virtual terminal password, use the following commands:

```
LINE VTY 0 4
LOGIN
PASSWORD [password]
```

➤ *Auxiliary Password*—The auxiliary password is used to control access to any auxiliary ports the router may have. Setting the auxiliary password will require all users connecting to the auxiliary port of the router (usually remote dial-in) to provide this password for access. To configure the auxiliary password, use the following commands:

```
LINE AUX 0
LOGIN
PASSWORD [password]
```

➤ *Console Password*—The console password is used to control access to the console port of the router. Setting the console password will require all users connecting to the router console to provide this password for access. To configure the console password, use the following commands:

```
LINE CON 0
LOGIN
PASSWORD [password]
```

For more information, see Chapter 8 of *CCNA Routing and Switching Exam Cram, Second Edition.*

Practice Test #4

Question 1

Which of the following protocols reside at the application layer of the OSI model? [Choose the four best answers]

❏ a. SMTP

❏ b. FTP

❏ c. Telnet

❏ d. WAIS

❏ e. PICT

❏ f. MIDI

❏ g. TIFF

Question 2

Consider the following display on your router console:

```
Router con0 is now available
    Press RETURN to get started.
    Router>ENABLE
    Password:
    Router#CONFIG T
    Enter configuration commands, one per line
 End with CTRL/Z.
    Router(config)#LINE AUX 0
    Router(config-line)#
```

Assuming you want to set a virtual terminal password, what would be the next command you would enter at the router prompt?

○ a. **PASSWORD LOGIN**

○ b. **SET LOGIN PASSWORD**

○ c. **LOGIN**

○ d. **SET LOGIN**

Question 3

Which of the following statements is true about connection-oriented and connectionless protocols?

○ a. Connection-oriented protocols feature the creation of a virtual circuit.

○ b. TCP is a connection-oriented protocol.

○ c. TCP is a connectionless protocol.

○ d. Connectionless protocols' overhead is more than connection-oriented protocols.

Question 4

The most common types of connections used in wide area networks (WANs) can be divided into digital communication, analog communication, and packet switching technology. Which of the following WAN technologies are considered packet switching technologies? [Choose the two best answers]

❑ a. ISDN

❑ b. POTS

❑ c. Frame relay

❑ d. X.25

Question 5

Consider the following access lists:

```
ACCESS-LIST 110 PERMIT TCP HOST 77.90.50.22
HOST 210.34.5.20 EQ 25
     ACCESS-LIST 110 DENY TCP 150.150.0 0
0.0.255.255 ANY EQ 21
     ACCESS-LIST 110 PERMIT ANY ANY
```

Based on this access list, which of the following statements are true? [Choose the three best answers]

❑ a. Host 210.34.5.20 will be able to access host 77.90.50.22 using SMTP.

❑ b. Host 77.90.50.22 will be able to access host 210.34.5.20 using SMTP.

❑ c. No computers on network 150.150.0.0 will be able to access any computers using FTP.

❑ d. No computers on network 150.150.0.0 will be able to access any computers using Telnet.

❑ e. All computers on the network will be able to communicate using Telnet.

❑ f. Only host 77.90.50.22 will be able to communicate using Telnet.

Question 6

You need to control the amount of traffic that your Novell NetWare servers are generating through their use of the Service Advertising Protocol. Your NetWare servers are sending out these "advertisements" every 60 seconds, and your Cisco routers are sending these updates throughout the internetwork at the same rate. You want to control the amount of Service Advertising Protocol traffic that your routers are passing. Which of the following access list types could you create to do this?

○ a.   IP standard access list

○ b.   IP extended access list

○ c.   IPX SAP access list

○ d.   Both a and b

Question 7

Encapsulation involves five steps. First, user information is converted to data. Second, data is converted to segments. Third, segments are converted to packets or datagrams. Fourth, packets and datagrams are converted to frames. What is the fifth step of encapsulation?

○ a.   Frames are sent over the physical medium.

○ b.   Frames are converted to bytes.

○ c.   Frames are converted to bits.

○ d.   Frames are converted to data.

Question 8

A router's motherboard contains the bootstrap program, the Power On Self Test (POST), and the operating system software (Cisco IOS). On which of the following core components of a Cisco router is a physical chip installed?

○ a.   Flash

○ b.   RAM

○ c.   ROM

○ d.   NVRAM

Question 9

Which of the following statements is true about accessing interface configuration mode on a Cisco router?

- ○ a. To access interface configuration mode, the **INTERFACE** command is used from any configuration mode.
- ○ b. To access interface configuration mode, the **INTERFACE** command is used from privileged mode.
- ○ c. To access interface configuration mode, the **INTERFACE** command is used from user mode.
- ○ d. To access interface configuration mode, the **INTERFACE** command is used from global configuration mode.

Question 10

What is the name of the command interpreter used by Cisco routers?

- ○ a. COMMAND.COM
- ○ b. CMD.EXE
- ○ c. EXEC
- ○ d. Supervisor

Question 11

Consider the following display on your Cisco router:

```
Router con0 is now available
    Press RETURN to get started.
    Router>
```

What mode is this router presently in?

- ○ a. User mode
- ○ b. Privileged mode
- ○ c. Executive mode
- ○ d. Global configuration mode

Question 12

You are configuring your Cisco router to use RIP. Which of the following is the correct command to enable RIP on the router and have it advertise network 131.200.0.0?

○ a.
```
ROUTER RIP
    NETWORK 131.200.0.0
```

○ b.
```
ENABLE RIP
    NETWORK 131.200.0.0
```

○ c.
```
ROUTER-RIP
    NETWORK 131.200.0.0
```

○ d.
```
RIP ROUTER
    NETWORK 131.200.0.0
```

Question 13

Which of the following are valid parameters used with the **CONFIGURE (CONFIG)** command? [Choose the four best answers]

❑ a. TERMINAL

❑ b. INTERFACE

❑ c. NETWORK

❑ d. HOST

❑ e. MEMORY

❑ f. OVERWRITE-NETWORK

❑ g. VIRTUAL-TERMINAL

Question 14

The TCP/IP configuration of one of the computers in your network is as follows:

```
IP address 208.100.50.90
    Subnet Mask 255.255.255.192
```

Based on this information, how many total bits are being used by subnet mask 255.255.255.192?

○ a. 16

○ b. 32

○ c. 24

○ d. 26

Question 15

Seated at the router console, you wish to change the setting for the clock. You are not sure of the command to use, but you recall it does begin with "CL". You decide to use the context-sensitive help feature to assist you. You enter the following command at the router prompt:

```
Router#CL ?
```

After entering this command, you receive the following result, as displayed by the router:

```
% Ambiguous command:  "CL "
```

Which of the following best describes the reason the router returned the result "% Ambiguous command: "CL "?

○ a. More than one command begins with the letters "CL".

○ b. You entered the wrong command. You should have entered "HELP CL" instead.

○ c. You entered the wrong command. You should have entered "HELP CL?" instead.

○ d. You entered the wrong command. Type "SHOW CL*" instead.

Question 16

Which of the following is the correct key(s) used to move the cursor forward one word in a Cisco router?

○ a. Ctrl+A

○ b. Ctrl+E

○ c. Esc+E

○ d. Ctrl+N

○ e. Esc+B

○ f. Tab

○ g. Esc+F

Question 17

The Cisco IOS includes a command history feature that can be used to store previously entered commands in a history buffer. The command history feature is enabled by default, but it can be disabled. Which of the following commands can be used to disable the command history feature on a Cisco router?

○ a. **DISABLE TERMINAL EDITING**

○ b. **TERMINAL EDITING DISABLE**

○ c. **TERMINAL EDITING OFF**

○ d. **TERMINAL NO EDITING**

Question 18

You are experiencing problems with the running configuration on your Cisco router. You have a backup copy of this configuration stored on a TFTP server. Which of the following is the correct Cisco IOS command that you could use to copy the configuration file from your TFTP server to the current running configuration of your router?

○ a. **COPY TFTP RUNNING-CONFIG**

○ b. **COPY TFTP RUNNING CONFIG**

○ c. **COPY TFTP TO RUNNING-CONFIG**

○ d. **COPY TFTP SERVER RUNNING-CONFIG**

Question 19

Which of the following are the correct key(s) used to return to more recent commands in the command history after recalling previous commands? [Choose the two best answers]

❑ a. Ctrl+A

❑ b. Ctrl+E

❑ c. Ctrl+P

❑ d. Ctrl+N

❑ e. Esc+B

❑ f. Down arrow

❑ g. Up arrow

Question 20

Which of the following protocols reside at the presentation layer of the OSI model? [Choose the three best answers]

❑ a. SMTP

❑ b. FTP

❑ c. Telnet

❑ d. WAIS

❑ e. PICT

❑ f. MIDI

❑ g. TIFF

Question 21

Your Cisco router console displays the following:

```
Router con0 is now available
      Press RETURN to get started.
      Router>enable
      Password:
      Router#
```

What command would you enter at this point to enter global configuration mode?

○ a. **CONFIG T**.

○ b. **CONFIG-T**.

○ c. **CONFIGURE GLOBAL**.

○ d. The router is already in global configuration mode.

Question 22

Which of the following best describes the use of the enable secret password in a Cisco router?

○ a. The enable secret password is used to control access to the Ethernet port on the router.

○ b. The enable secret password is used to control access to the privileged mode of the router and is not used if the enable password is also enabled.

○ c. The enable secret password is used to control access to any auxiliary ports the router may have.

○ d. The enable secret password is used to control access to the privileged mode of the router and takes precedence over the enable password if both are enabled.

○ e. The enable secret password is used to control access to the console port of the router and takes precedence over the enable password if both are enabled.

○ f. The enable secret password is used to control access to the user mode of the router.

Question 23

You want to configure a welcome message that will be displayed by your Cisco router to users when they log in. You enter the following from your router console:

```
Router>ENABLE
    Password:
    Router#BANNER MOTD #
                 ^
    % Invalid input detected at '^' marker.
    Router#
```

Which of the following is the most likely reason this command failed?

○ a. To use the **BANNER MOTD** command, you cannot use a pound sign (#) at the end.

○ b. You entered the **BANNER MOTD** command in privileged mode, and you must be in user mode to issue the **BANNER MOTD** command.

○ c. You entered the **BANNER MOTD** command in user mode, and you must be in privileged mode to issue the **BANNER MOTD** command.

○ d. You entered the **BANNER MOTD** command in privileged mode, and you must be in global configuration mode to issue the **BANNER MOTD** command.

Question 24

You have been assigned a Class B IP address for use in your network. You will use subnet mask 255.255.224.0.

Based on this information, how many total bits are available for host IDs on this network?

○ a. 16

○ b. 19

○ c. 13

○ d. 18

Question 25

Seated at your router console, you create the following access list:

```
ACCESS-LIST 88 PERMIT 200.18.100.5 0.0.0.255
```

You want to apply this access list to the outgoing serial interface number 1 of your router. Which of the following commands will you enter to perform this operation?

○ a. **ROUTER# IP ACCESS-GROUP 88 OUT**

○ b. **ROUTER#IP-ACCESS-GROUP 88 OUT**

○ c. **ROUTER(config-if)#ACCESS-GROUP 10 OUT**

○ d. **ROUTER(config-if)#IP ACCESS-GROUP 88 OUT**

○ e. **ROUTER(config-if)#IP ACCESS GROUP 88 OUT**

○ f. **ROUTER>IP ACCESS-GROUP 88 OUT**

Question 26

Seated at your router console, you create the following access list:

```
ACCESS-LIST 155 PERMIT TCP HOST 77.90.50.22
 HOST 210.34.5.20 EQ 25
    ACCESS-LIST 155 DENY TCP
150.150.0 0 0.0.255.255 ANY EQ 21
    ACCESS-LIST 155 PERMIT ANY ANY
```

You want to apply this access list to the outgoing serial interface number 0 of your router. You enter the interface configuration mode of your router and access serial interface 1.

Which of the following commands will you enter to perform this operation?

○ a. **IP ACCESS-GROUP 10 OUT**

○ b. **IP ACCESS-GROUP 155 OUT**

○ c. **ACCESS-GROUP 155 OUT**

○ d. **IP ACCESS-GROUP SO OUT**

Question 27

Which layer of the OSI model is responsible for flow control, multiplexing, and end-to-end communication?

○ a. Application

○ b. Presentation

○ c. Session

○ d. Transport

○ e. Network

○ f. Data-link

○ g. Physical

Question 28

A standard IPX access list can be created to filter the IPX traffic that passes through a router. When specifying the source or destination IPX network in a standard IPX access list, a "-1" can be used. Which of the following best describes what "-1" means when it is used as the source or destination network in a standard IPX access list?

○ a. "-1" in the source or destination network of a standard IPX access list means "all networks."

○ b. "-1" in the source or destination network of a standard IPX access list means "all networks that have a "1" in their network ID.

○ c. "-1" in the source or destination network of a standard IPX access list means "this network."

○ d. "-1" in the source or destination network of a standard IPX access list means "no networks."

Question 29

You have configured several IP access lists on your Cisco router. Which of the following commands can you use to see a list of all the IP access lists that are currently in use on your router, as well as what lines in those access lists were matched by the traffic that passed through the router?

- ○ a. **SHOW IP ACCESS LISTS**
- ○ b. **SHOW IP INTERFACE**
- ○ c. **SHOW IP ACCESS-LISTS**
- ○ d. **SHOW IP INTERFACE E0**

Question 30

Which of the following protocols reside at the data-link layer of the OSI model? [Choose the five best answers]

- ❏ a. HDLC
- ❏ b. SDLC
- ❏ c. NFS
- ❏ d. SLIP
- ❏ e. PPP
- ❏ f. RPC
- ❏ g. LAPB

Question 31

You have configured an IPX SAP access list on your Cisco router. The access list number is 1091. You want to see what router interface this IPX SAP access list has been applied to. Which of the following commands can you use to view this information?

- ○ a. **SHOW IPX INTERFACE**
- ○ b. **SHOW IPX-INTERFACE**
- ○ c. **SHOW SAP INTERFACE**
- ○ d. **SHOW SAP-INTERFACE**

Question 32

The most common form of WAN connectivity utilizes telecommunication technology. Which of the following best describes CPE as it is used in WAN communication?

○ a. CPE, or customer premises equipment, is the collection of communication devices (telephones, modems, and so forth) that exists at the customer's location.

○ b. CPE, or central premises equipment, is the telephone company's office that acts as the central communication point for the customer.

○ c. CPE, or circuit provider equipment, is the cabling that extends from the DEMARC to the telephone company's office.

○ d. None of these choices is correct.

Question 33

Frame relay is a packet switching WAN technology that utilizes logical or virtual circuits to form a connection. Frame relay networks utilize Local Management Interface (LMI) signaling types that are used to manage a frame relay connection. Which of the following statements correctly describes the LMI signaling type supported by Cisco routers?

○ a. Cisco routers support only the CISCO LMI signaling type.

○ b. Cisco routers support three LMI signaling types: CISCO, ANSI, and Q933A. The Cisco default LMI type is CISCO.

○ c. Cisco routers support three LMI signaling types: CISCO, ANSI, and Q933A. The Cisco default LMI type is ANSI.

○ d. Cisco routers support three LMI signaling types: CISCO, ANSI, and Q933A. The Cisco default LMI type is Q933A.

Question 34

ISDN, or Integrated Services Digital Network, is a communications standard that uses digital telephone lines to transmit voice, data, and video. ISDN is a dial-up service that is used on demand, and its protocol standards are defined by the International Telecommunication Union (ITU). Which of the following statements is true about ISDN protocol standards?

○ a. Protocols that begin with the letter *I* define ISDN standards on the existing telephone network.

○ b. Protocols that begin with the letter *E* define ISDN standards on the existing telephone network.

○ c. Protocols that begin with the letter *E* specify ISDN switching and signaling standards.

○ d. Protocols that begin with the letter *Q* specify ISDN concepts, interfaces, and terminology.

Question 35

The Cisco IOS includes a command history feature that can be used to store previously entered commands in a history buffer. Which of the following commands would you enter to set the number of command lines the command history buffer will hold?

○ a. **TERMINAL HISTORY SIZE**

○ b. **CHANGE TERMINAL HISTORY**

○ c. **HISTORY SIZE**

○ d. **HISTORY BUFFER SIZE**

Question 36

Consider the following IPX address:

```
5F.1111.2222.3333
```

Based on this information, what is the node ID of this address?

○ a. 5F.1111

○ b. 1111.2222.3333

○ c. 1111.2222

○ d. 5F

Question 37

Which of the following is the correct command and prompt used to enter the CONFIGURE OVERWRITE-NETWORK?

- ○ a. **ROUTER> CONFIG O**
- ○ b. **ROUTER#CONFIG O**
- ○ c. **ROUTER(config)#CONFIG O**
- ○ d. **ROUTER(CONFIG-IF)#CONFIG O**

Question 38

Which of the following statements is true regarding flow control methods?

- ○ a. Windowing, buffering, and source-quench messages are used in flow control.
- ○ b. Buffering, multiplexing, and windowing are used in flow control.
- ○ c. Source-quench messages, buffering, and multiplexing are used in flow control.
- ○ d. None of these choices is correct.

Question 39

Novell NetWare supports a number of different encapsulation methods or frame types used to encapsulate IPX packets. The frame type used by NetWare computers must be the same as the one used by a Cisco router, or they will not be able to communicate. Novell NetWare and Cisco both use different names to describe the same frame type. Which of the following statements are true about the frame types used by Novell and Cisco? [Choose the three best answers]

- ❑ a. The TOKEN RING_SNAP frame type used by Novell is called SAP in Cisco.
- ❑ b. The ETHERNET 802.2 frame type used by Novell is called SNAP in Cisco.
- ❑ c. The ETHERNET_snap frame type used by Novell is called SAP in Cisco.
- ❑ d. The ETHERNET 802.3 frame type used by Novell is called NOVELL-ETHER in Cisco and is the default frame type.
- ❑ e. The TOKEN RING frame type used by Novell is called TOKEN in Cisco.
- ❑ f. The ETHERNET II frame type used by Novell is called ARPA in Cisco.

Question 40

A dynamic routing table is one that has its entries entered automatically by the router through the use of a routing protocol. A dynamic routing protocol provides for automatic discovery of routes and eliminates the need for static routes. Routing protocols generally fall into one of two categories: link state and distance vector. Which of the following is considered a link state routing protocol?

○ a. RIP

○ b. OSPF

○ c. EIGRP

○ d. IGRP

Question 41

Which layer of the OSI model makes use of logical addresses to select the best path a packet will take to its destination?

○ a. Application

○ b. Presentation

○ c. Session

○ d. Transport

○ e. Network

○ f. Data-link

○ g. Physical

Question 42

A Cisco router's operating system software is known as the Cisco IOS. In what two places can the Cisco IOS be permanently stored in a Cisco router? [Choose the two best answers]

❑ a. RAM

❑ b. ROM

❑ c. NVRAM

❑ d. Flash

Question 43

Consider the following access list:

```
ACCESS-LIST 10 PERMIT 131.200.0.0 0.0.255.255
```

Based on this information, which of the following statements best describes the result of this access list?

○ a.　This access list will allow traffic from host 131.200.0.0.

○ b.　This access list will allow traffic from host 131.200.0.0 to enter network 255.255.0.0.

○ c.　This access list will allow traffic from all hosts on network 131.200.0.0.

○ d.　This is not an access list.

Question 44

You manage the Cisco routers for your company, and your company maintains seven routers in all. After modifying the running configuration on one of these routers, you realize that you have configured the wrong router. All of the new settings that you have changed on this router must be removed. Which of the following commands could you use to return this router to the configuration it was using before you made changes to its running configuration?

○ a.　Issue the **COPY RUNNING-CONFIG STARTUP CONFIG** command and restart the router.

○ b.　Issue the **WRITE STARTUP-CONFIG RUNNING-CONFIG** command and restart the router.

○ c.　Issue the **WRITE RUNNING-CONFIG STARTUP-CONFIG** command and restart the router.

○ d.　Issue the **COPY STARTUP-CONFIG RUNNING-CONFIG** command and restart the router.

Question 45

Which layer of the OSI model defines the voltage levels and physical components of a network?

○ a. Application

○ b. Presentation

○ c. Session

○ d. Transport

○ e. Network

○ f. Data-link

○ g. Physical

Question 46

Both RIP and IGRP are dynamic routing protocols used to automatically update entries in a routing table. Which of the following statements are true about IGRP and RIP? [Choose the three best answers]

❑ a. Both IGRP and RIP are distance vector protocols.

❑ b. To select the best route, RIP only uses hop count.

❑ c. IGRP considers hop count, delay, bandwidth, and reliability when determining the best path.

❑ d. The maximum hop count used in IGRP is 15.

❑ e. The maximum hop count used in RIP is 255.

❑ f. Both IGRP and RIP are link state routing protocols.

❑ g. RIP considers hop count, delay, bandwidth, and reliability when determining the best path.

Question 47

You need to configure a static route on your Cisco router, and you want the router to send all packets destined for network 190.190.0.0 255.255.0.0 to router 10.11.12.1. Which of the following is the correct command to create a static route that will perform this operation?

○ a. **IP-ROUTE 190.190.0.0 255.255.0.0 10.11.12.1**

○ b. **ROUTE 190.190.0.0 255.255.0.0 10.11.12.1**

○ c. **IP ROUTE 10.11.12.1 255.255.0.0 190.190.0.0**

○ d. **IP ROUTE 190.190.0.0 255.255.0.0 10.11.12.1**

Question 48

Your colleague, Don, is having trouble logging into privileged mode on his Cisco router and asks for your help. Don tells you that he configured the router to use both an enable password *and* an enable secret password. The enable password is "FULLER", and the enable secret password is "KATE". Seated at the router, you attempt to access the router's privileged mode. Based on the above configuration information, what password should you attempt to use to gain access to the privileged mode of this router?

- ○ a. No password is needed, because neither the enable nor enable secret passwords control access to privileged mode of a router.

- ○ b. Use the password "FULLER" to access privileged mode.

- ○ c. Use the password "KATE" to access privileged mode.

- ○ d. Both "KATE" and "FULLER" will work. It does not matter which one you use.

Question 49

Which of the following statements best describes the use of the **CONFIGURE MEMORY (CONFIG MEM)** command?

- ○ a. **CONFIGURE MEMORY** is used to enter configuration commands into the router from the console port or through Telnet.

- ○ b. **CONFIGURE MEMORY** is used to copy the configuration file from a TFTP server into the router's RAM.

- ○ c. **CONFIGURE MEMORY** is used to copy a configuration file into NVRAM from a TFTP server.

- ○ d. **CONFIGURE MEMORY** is used to execute the configuration stored in NVRAM and will copy the startup configuration to the running configuration.

Question 50

Seated at your Cisco router, you wish to rerun the initial system configuration dialog. Which of the following commands would you enter to perform this operation?

- ○ a. Run **SETUP** from user mode

- ○ b. Run **INSTALL** from user mode

- ○ c. Run **SETUP** from privileged mode

- ○ d. Run **SYS-CONFIG** from privileged mode

Question 51

You are configuring your Cisco router to use IGRP. Which of the following is the correct command to enable IGRP on the router and have it advertise network 200.200.200.0 in autonomous system number 15?

○ a.
```
ROUTER IGRP
    NETWORK 200.200.200.0 15
```

○ b.
```
ROUTER IGRP 15
    NETWORK 200.200.200.0
```

○ c.
```
ROUTER-IGRP 15
    NETWORK 200.200.200.0
```

○ d.
```
ROUTER IGRP AS 15
    NETWORK 200.200.200.0
```

Question 52

Which of the following Exec modes is used to modify and view the configuration of the router, as well as set IOS parameters?

○ a. User mode

○ b. Privileged mode

○ c. Both user mode and privileged mode

○ d. Exec mode

Question 53

You decide to implement a Cisco switch in your network. You want to implement a switching method that offers error checking of packets before they are forwarded by your switch. Which of the following switching methods will you use in this situation?

○ a. Cut-through switching.

○ b. Either store-and-forward switching or cut-through switching.

○ c. Store-and-forward switching.

○ d. None of these switching methods provides error checking.

Question 54

Which of the following protocols provides a logical network layer address? [Choose the two best answers]

❏ a. IPX

❏ b. SPX

❏ c. TCP

❏ d. IP

Question 55

Key components of WAN communication are data terminal equipment (DTE) and data communications equipment (DCE) devices. Which of the following best describes a DTE device?

○ a. DTE devices are modems used when communicating over analog phone lines.

○ b. A DTE device is a CSU/DSU or channel service unit/data service unit used when communication is over digital lines.

○ c. A DTE device is usually the router that acts as the gateway to the local area network (LAN).

○ d. DTE devices are the communication devices (telephones, modems, and so forth) that exist at the customer's location.

Question 56

Consider the following access list:

```
ACCESS-LIST 10 PERMIT 131.200.0.0 0.0.255.255
```

Based on this information, what type of access list is this?

○ a. This is a standard IPX access list.

○ b. This is an extended IP access list.

○ c. This is a standard IP access list.

○ d. This is an extended IPX access list.

Question 57

You are configuring IP on serial interface S0 of your Cisco router. You specify the following at the router console:

```
Configuring interface Serial0:
    Is this interface in use? [no]: y
    Configure IP on this interface? [no]: y
    Configure IP unnumbered on this interface?
 [no]: n
    IP address for this interface:
131.200.15.1
    Number of bits in subnet field [0]: 5
```

Based on this information, which of the following statements is true?

○ a. This is a Class C address, and the subnet mask is 255.255.252.0.

○ b. This is a Class B address, and the subnet mask is 255.255.0.0.

○ c. This is a Class B address, and the subnet mask is 255.255.248.0.

○ d. This is a Class A address, and the subnet mask is 255.255.224.0.

Question 58

ISDN, or Integrated Services Digital Network, is a communications standard that uses digital telephone lines to transmit voice, data, and video. ISDN is a dial-up service that is used on demand. The bandwidth provided by ISDN can be divided into two categories: Primary Rate ISDN (PRI) and Basic Rate ISDN (BRI). Which of the following statements is true about Basic Rate ISDN?

○ a. Basic Rate ISDN (also known as 2B+1E) comprises two B channels that are 64Kbps each and one 16Kbps E channel.

○ b. Basic Rate ISDN (also known as 2B+1D) comprises two B channels that are 16Kbps each and one 64Kbps D channel.

○ c. Basic Rate ISDN (also known as 2B+1D) comprises two B channels that are 64Kbps each and one 16Kbps D channel.

○ d. Basic Rate ISDN (also known as 2D+1B) comprises two D channels that are 64Kbps each and one 16Kbps B channel.

Question 59

A Cisco router's startup sequence is divided into three basic operations. First, the bootstrap program in ROM executes the POST, which performs a basic hardware level check. What is the next step in this startup sequence?

○ a. The router's configuration file is loaded into memory from NVRAM.

○ b. The boot loader program executes.

○ c. The router's configuration file is loaded into memory from ROM.

○ d. The Cisco IOS (operating system software) is loaded into memory per instructions from the boot system command.

Question 60

Which of the following statements are true about hierarchical and flat addressing? [Choose the two best answers]

❏ a. The flat addressing used in IP provides a more organized environment than hierarchical addressing.

❏ b. The hierarchical addressing used in IP provides a more organized environment than flat addressing.

❏ c. The flat addressing of MAC addresses enables complex networks to be logically grouped and organized.

❏ d. The hierarchical addressing of IP and IPX enable complex networks to be logically grouped and organized.

Question 61

Which of the following access lists will allow all hosts on network 200.200.200.0 to have FTP access to network 151.120.0.0?

○ a. **ACCESS-LIST 102 PERMIT TCP 200.200.200.0 0.0.0.255 151.120.0.0 0.0 255.255 EQ 23**

○ b. **IP ACCESS-LIST 101 PERMIT TCP 200.200.200.0 0.0.0.255 151.120.0.0 0.0.255.255 EQ 21**

○ c. **ACCESS-LIST 99 PERMIT TCP 200.200.200.0 0.0.0.255 151.120.0.0 0.0.255.255 EQ 21**

○ d. **ACCESS-LIST 101 PERMIT TCP 200.200.200.0 0.0.0.255 151.120.0.0 0.0.255.255 EQ 21**

Question 62

You manage a network that uses TCP/IP as the only protocol and is divided into multiple subnets. The IP configuration of one of your subnets is as follows:

```
Broadcast Address = 129.23.98.63 on interface E0
    Subnet Mask = 255.255.255.224
```

Based on this information, what is the valid host ID for the nodes in this subnet?

○ a. 129.23.98.32 to 129.23.98.63

○ b. 129.23.98.32 to 129.23.98.62

○ c. 129.23.98.33 to 129.23.98.63

○ d. 129.23.98.33 to 129.23.98.62

Question 63

Which of the following best describes the function and contents of Read-Only Memory (ROM) in a Cisco router?

○ a. ROM is a physical chip installed on a router's motherboard that contains the bootstrap program, the POST, and the operating system software (Cisco IOS).

○ b. ROM is used by Cisco routers to store the startup configuration.

○ c. ROM is an erasable, programmable memory area that contains the Cisco operating system software (IOS).

○ d. ROM is used as the router's main working area and contains the running configuration.

Question 64

As the network administrator, you want to require a password to be entered by any user connecting to your Cisco router via Telnet. You want the password to be "JETS". Which of the following is the correct Cisco IOS command to perform this operation?

○ a.
```
LINE VTY 0 4
    LOGIN
    PASSWORD JETS
```

○ b.
```
LINE CON 0 4
    LOGIN
    PASSWORD JETS
```

○ c.
```
LINE VTY 0 4
    LOGIN JETS
```

○ d.
```
LINE AUX 0 4
    PASSWORD JETS
```

Question 65

To control the amount of SAP advertisement traffic that is passed by your Cisco router, you create the following IPX SAP access list:

```
ACCESS LIST 1000 DENY -1  7
    ACCESS LIST 1000 PERMIT -1
```

You want to apply this access list to the outgoing serial interface number 0 of your router. You enter the interface configuration mode of your router and access serial interface 0.

Which of the following commands will you enter to perform this operation?

○ a. **IPX OUTPUT-SAP-FILTER 1000**

○ b. **OUTPUT-SAP-FILTER 1000**

○ c. **IPX OUTPUT SAP FILTER 1000**

○ d. **IPX ACCESS-GROUP 1000 OUT**

Question 66

You observe the following configuration on one of the computers in your TCP/IP network:

```
IP address: 222.111.101.200
    Subnet Mask 255.255.255.255
    Default Gateway: 222.111.101.254
```

Based on this information, which of the following statements are true? [Choose the three best answers]

❑ a. The IP address is a Class C address.

❑ b. The IP address is a Class A address.

❑ c. The network ID of IP address 222.111.101.200 is 222.111.0.0.

❑ d. The network ID of IP address 222.111.101.200 is 222.111.101.0.

❑ e. The host ID of IP address 222.111.101.200 is 101.200.

❑ f. The host ID of IP address 222.111.101.200 is 200.

Question 67

Which layer of the OSI model is responsible for creating, maintaining, and ending a communication session?

○ a. Application

○ b. Presentation

○ c. Session

○ d. Transport

○ e. Network

○ f. Data-link

○ g. Physical

Question 68

Seated at the console of your Cisco router, you wish to display the router's startup configuration. Which of the following Cisco IOS commands could you enter to perform this operation? [Choose the two best answers]

❑ a. **ROUTER# SHOW STARTUP-CONFIG**

❑ b. **ROUTER# SH STAR**

❑ c. **ROUTER>SHOW STARTUP-CONFIG**

❑ d. **ROUTER#SHOW STARTUP CONFIG**

❑ e. **ROUTER>SHOW STARTUP-CONFIG**

❑ f. **ROUTER>SH STAR**

❑ g. **ROUTER#DISPLAY STARTUP**

Question 69

Which of the following Cisco IOS commands is used to execute the configuration stored in NVRAM and will copy the startup configuration to the running configuration?

○ a. **CONFIGURE TERMINAL (CONFIG T)**

○ b. **CONFIGURE OVERWRITE-NETWORK (CONFIG O)**

○ c. **CONFIGURE NETWORK (CONFIG NET)**

○ d. **CONFIGURE MEMORY (CONFIG MEM)**

Question 70

Which of the following correctly states the use of Ctrl+N at the command line in a Cisco router?

○ a. Ctrl+N will move the cursor to the beginning of the command line.

○ b. Ctrl+N will recall previous commands in the command history.

○ c. Ctrl+N will move the cursor back one word.

○ d. Ctrl+N will return to more recent commands in the command history after recalling previous commands.

Answer Key #4

1. a, b, c, d
2. c
3. a, b
4. c, d
5. b, c, e
6. c
7. c
8. c
9. d
10. c
11. a
12. a
13. a, c, e, f
14. d
15. a
16. g
17. d
18. a

19. d, f
20. e, f, g
21. a
22. d
23. d
24. c
25. d
26. b
27. d
28. a
29. c
30. a, b, d, e, g
31. a
32. a
33. b
34. b
35. a
36. b

37. b
38. a
39. d, e, f
40. b
41. e
42. b, d
43. c
44. d
45. g
46. a, b, c
47. d
48. c
49. d
50. c
51. b
52. b
53. c
54. a, d

55. c
56. c
57. c
58. c
59. d
60. b, d
61. d
62. d
63. a
64. a
65. a
66. a, d, f
67. c
68. a, b
69. d
70. d

Question 1

The correct answers are a, b, c, and d. The OSI model defines standards used in networking and comprises a seven-layer model. The application layer of the OSI model functions as the "window" for applications to access network resources and is responsible for verifying the identity of the destination node. Application layer protocols include Telnet, File Transfer Protocol (FTP), Simple Mail Transfer Protocol (SMTP), World Wide Web (WWW), Electronic Data Interchange (EDI), and Wide Area Information Server (WAIS).

Answers e, f, and g are incorrect, because all are presentation layer protocols. The presentation layer functions as a translator and is responsible for text and data syntax translation, such as Extended Binary-Coded Decimal Interchange Code (EBCDIC, used in IBM systems) to American Standard Code for Information Interchange (ASCII, used in PC and most computer systems). Other presentation layer protocols include PICTure (PICT, Apple computer picture format), Tagged Image File Format (TIFF), Joint Photographic Experts Group (JPEG), Musical Instrument Digital Interface (MIDI), Motion Picture Expert Group (MPEG), and Quick Time (audio/video application).

For more information, see Chapter 4 of *CCNA Routing and Switching Exam Cram, Second Edition.*

Question 2

The correct answer is c. Based on the display, the next command to issue to set an auxiliary password would be "**LOGIN**". Answers a, b, and d are all incorrect uses of the **LOGIN** command and are, therefore, incorrect. Cisco routers utilize passwords to provide security access to a router. Passwords can be set for controlling access to privileged mode, for controlling access to remote sessions via Telnet, or for controlling access through the auxiliary port. The auxiliary password is used to control access to any auxiliary ports the router may have. Setting the auxiliary password will require all users connecting to the auxiliary port of the router (usually remote dial-in) to provide this password for access. To configure the auxiliary password, use the following commands:

```
LINE AUX 0
    LOGIN
    PASSWORD [password]
```

For more information, see Chapter 8 of *CCNA Routing and Switching Exam Cram, Second Edition.*

Question 3

The correct answers are a and b. User Datagram Protocol (UDP) is a transport layer protocol that is connectionless. Answer c is incorrect, because TCP is a connection-oriented protocol. Answer d is incorrect, because connection-oriented protocols have more overhead than connectionless protocols. Connectionless protocols feature:

➤ *No Sequencing or Virtual Circuit Creation*—Connectionless protocols do not sequence packets or create virtual circuits.

➤ *No Guarantee of Delivery*—Packets are sent as datagrams, and delivery is not guaranteed by connectionless protocols. When using a connectionless protocol like UDP, the guarantee of delivery is the responsibility of higher layer protocols.

➤ *Less Overhead*—Because connectionless protocols do not perform any of the above services, overhead is less than connection-oriented protocols. Transmission Control Protocol (TCP) is a transport layer protocol that is connection-oriented. Connection-oriented protocols feature:

➤ *Reliability*—A communication session is established between hosts before sending data. This session is considered a virtual circuit.

➤ *Sequencing*—When a connection-oriented protocol like TCP sends data, it numbers each segment, so the destination host can receive the data in the proper order.

➤ *Guaranteed Delivery*—Connection-oriented protocols utilize error checking to guarantee packet delivery.

➤ *More Overhead*—Because of the additional error checking and sequencing responsibilities, connection-oriented protocols require more overhead than connectionless protocols.

For more information, see Chapter 9 of *CCNA Routing and Switching Exam Cram, Second Edition.*

Question 4

The correct answers are c and d. Frame relay and X.25 are packet switching technologies. Packet switching is a process of sending data over wide area network (WAN) links through the use of logical circuits, where data is encapsulated into packets and routed by the service provider through various switching points.

Answer a is incorrect, because Integrated Services Digital Network (ISDN) is a communications standard that uses digital communication lines. Answer b is incorrect, because plain old telephone service (POTS) is a standard analog phone line and is used as a dial-up service using modems.

For more information, see Chapter 11 of *CCNA Routing and Switching Exam Cram, Second Edition.*

Question 5

The correct answers are b, c, and e. The first access list, **ACCESS-LIST 110 PERMIT IP HOST 77.90.50.22 HOST 210.34.5.20 EQ 25**, specifies that SMTP (port **25**) traffic is permitted from host **77.90.50.22** to host **210.34.5.20**. The second access list, **ACCESS-LIST 110 DENY TCP 150.150.0 0 0.0.255.255 ANY EQ 21**, specifies that all hosts on network **150.150.0.0** are to be denied FTP traffic to all networks (**ANY**). The third access list, **AC-CESS-LIST 110 PERMIT ANY ANY**, specifies that any hosts that have not matched a previous line are permitted to send all traffic.

Answer a is incorrect, because the source and destination are in the wrong order. The syntax for an extended Internet Protocol (IP) access list specifies the source first, then the destination. Answer d is incorrect, because the port specified is 21, which is FTP. Telnet uses port 23. Answer f is incorrect, because the port specified is 25, which is SMTP. FTP uses port 21. The syntax for creating an extended IP access list is as follows:

```
ACCESS-LIST [access-list-number] [permit|deny] [protocol] [source]
    [wildcard mask] [destination] [wildmask] [operator][port]
```

➤ *ACCESS-LIST*—This is the command to denote an access list.

➤ *[access-list-number]*—This can be a number between 100 and 199, and is used to denote an extended IP access list.

➤ *[permit|deny]*—This will allow (permit) or disallow (deny) traffic specified in this access list.

➤ *[protocol]*—This is used to specify the protocol that will be filtered in this access list. Common values are TCP, UDP, and IP. (The use of IP will denote all IP protocols.)

➤ *[source]*—This specifies the source IP addressing information.

➤ *[wildcard mask]*—This can be optionally applied to further define the source. A wildmask can be used to control access to an entire IP network

ID rather than a single IP address. A wildcard mask will use the number 255 to mean "any" and the number 0 to mean "must match exactly." The terms *host* and *any* may also be used here to more quickly specify a single host or an entire network. The **HOST** keyword is the same as the wildcard mask 0.0.0.0; the **ANY** keyword is the same as the wildcard mask 255.255.255.255.

➤ *[destination]*—This specifies the destination IP addressing information.

➤ *[wildmask]*—This can be optionally applied to further define the destination IP addressing.

➤ *[operator]*—This can be optionally applied to define how to interpret the value entered in the **[port]** section of the access list. Common values are equal to the port specified (**EQ**), less than the port number specified (**LT**), and greater than the port number specified (**GT**).

➤ *[port]*—This specifies the port number this access list will act on. Common port numbers include 21 (FTP), 23 (TELNET), 25 (SMTP), 53 (DNS), 69 (TFTP), and 80 (HTTP).

For more information, see Chapter 13 of *CCNA Routing and Switching Exam Cram, Second Edition.*

Question 6

The correct answer is c. An IPX SAP filter can be applied to a router's interface to filter the Service Advertising Protocol (SAP) advertisement traffic passed by the router. Novell NetWare servers will broadcast a list of their available resources every sixty seconds. This is known as a SAP advertisement. Cisco routers will record this information into their own SAP table for advertisement to other segments. By filtering the amount of SAP traffic with an IPX SAP access list, network traffic can be controlled and/or reduced. Answers a, b, and d are incorrect, because neither a standard nor extended IP access list can filter SAP advertisement traffic.

For more information, see Chapter 13 of *CCNA Routing and Switching Exam Cram, Second Edition.*

Question 7

The correct answer is c. The OSI model defines a layered approach to network communication. Each layer of the OSI model adds its own layer-specific information and passes it on to the next layer until it leaves the computer and goes out onto the network. This process is known as encapsulation, and it involves five basic steps:

1. User information is converted to data.

2. Data is converted to segments.

3. Segments are converted to packets or datagrams.

4. Packets and datagrams are converted to frames.

5. Frames are converted to bits.

Since the only operation performed in the fifth step of encapsulation is frames-to-bits, all of the other answers are incorrect.

For more information, see Chapter 3 of *CCNA Routing and Switching Exam Cram, Second Edition.*

Question 8

The correct answer is c. Read-Only Memory (ROM) is a physical chip installed on a router's motherboard that contains the bootstrap program, the Power On Self Test (POST), and the operating system software (Cisco IOS). The internal components of Cisco routers are comparable to that of conventional PCs.

Answer a is incorrect, because Flash is an erasable, programmable memory area that contains the Cisco operating system software (IOS). Answer b is incorrect, because RAM is used as the router's main working area and contains the running configuration. All information in RAM dissipates when the router is turned off. Answer d is incorrect, because Non-Volatile RAM (NVRAM) is used by Cisco routers to store the startup or running configuration.

For more information, see Chapter 4 of *CCNA Routing and Switching Exam Cram, Second Edition.*

Question 9

The correct answer is d. Interface configuration mode is used to set interface-specific parameters on a Cisco router. To access interface configuration mode, the **INTERFACE** command is used. When issuing the **INTERFACE** command, it must be followed by the interface ID that is to be accessed. (**S0** for Serial port 0, **S1** for Serial port 1, **E0** for Ethernet port 0, and so forth.)

Because the **INTERFACE** command must be issued from global configuration mode, answers a, b, and c are incorrect. When a router is in global configuration mode, the prompt displays "**config**" in parentheses **(Router(config)#)**. The **SHOW INTERFACE** command is used to display the configuration information of all interfaces. If the **SHOW INTERFACE** command is issued with a specific port (that is, **SHOW INTERFACE E0**), then only the configuration of that interface will be displayed. The **SHOW INTERFACE** command can be issued from user or privileged mode

For more information, see Chapter 8 of *CCNA Routing and Switching Exam Cram, Second Edition.*

Question 10

The correct answer is c. The Cisco IOS includes a command interpreter called **EXEC**, which is used to interpret or execute the commands entered into a Cisco router by an operator. Answer a is incorrect, because **COMMAND.COM** is the DOS command interpreter and is not used by Cisco routers. Answer b is incorrect, because **CMD.EXE** is the command interpreter used by Windows NT. Answer d is incorrect, because Supervisor is not the command interpreter used by Cisco routers.

For more information, see Chapter 8 of *CCNA Routing and Switching Exam Cram, Second Edition.*

Question 11

The correct answer is a. The router is in user mode. The Cisco IOS uses a command interpreter known as **EXEC**, which contains two primary modes: user mode and privileged mode. User mode displays basic router system information and connects to remote devices. The commands in user mode are limited in capability. In user mode, the router prompt is followed by an angle bracket **(ROUTER>)**. Answer b is incorrect, because privileged mode is used to modify and view the configuration of the router, as well as to set IOS parameters. Privileged mode also contains the **CONFIGURE** command, which is used to access other configuration modes, such as global configuration mode and interface mode. In privileged mode, the router prompt is followed by a pound sign **(ROUTER#)**. To access privileged mode, enter the **ENABLE** command from user mode. Answers c and d are incorrect, because global configuration mode is part of privileged mode, and cannot be accessed directly from user mode.

For more information, see Chapter 8 of *CCNA Routing and Switching Exam Cram, Second Edition.*

Question 12

The correct answer is a. The **ROUTER RIP NETWORK 131.200.0.0** command will enable **RIP** on the router and have it advertise network **131.200.0.0**. Answer b is incorrect, because **ENABLE RIP** is not a valid Cisco IOS command. Answer c is incorrect, because the **ROUTER RIP** command does not have a hyphen between **ROUTER** and **RIP**. Answer d is incorrect, because **RIP ROUTER** is not a valid Cisco IOS command. RIP is a distance vector routing protocol used to automatically update entries in a routing table. RIP routers update their routing tables by broadcasting the contents of their routing table every 30 seconds. To enable RIP on a Cisco router, the **ROUTER RIP** command is used. The syntax for the **ROUTER RIP** command is as follows:

```
ROUTER RIP
    NETWORK [network id]
```

The **ROUTER RIP** command enables **RIP** on the router. The **NETWORK** command specifies the network ID of the route that will be advertised by the router.

For more information, see Chapter 11 of *CCNA Routing and Switching Exam Cram, Second Edition.*

Question 13

The correct answers are a, c, e, and f. The **CONFIGURE** command, executed in privileged mode, is used to enter the configuration mode of a Cisco router. None of the other answers are valid parameters used with the **CONFIGURE** command, therefore, answers b, d, and g are incorrect. The four parameters used with the **CONFIGURE** command are **TERMINAL**, **MEMORY**, **OVERWRITE-NETWORK**, and **NETWORK**. The following list outlines the functions of these parameters:

➤ *CONFIGURE TERMINAL (CONFIG T)*—This is used to enter configuration commands into the router from the console port or through Telnet.

➤ *CONFIGURE NETWORK (CONFIG NET)*—This is used to copy the configuration file from a TFTP server into the router's RAM.

➤ *CONFIGURE OVERWRITE-NETWORK (CONFIG O)*—This is used to copy a configuration file into NVRAM from a TFTP server. Use of the **CONFIGURE OVERWRITE-NETWORK** will not alter the running configuration.

➤ *CONFIGURE MEMORY (CONFIG MEM)*—This is used to execute the configuration stored in NVRAM. It will copy the startup configuration to the running configuration. **HOST, VIRTUAL-TERMINAL,** and **INTERFACE** are not valid **CONFIGURE** command parameters.

For more information, see Chapter 8 of *CCNA Routing and Switching Exam Cram, Second Edition.*

Question 14

The correct answer is d. The subnet mask 255.255.255.192 uses a total of 26 bits. An IP address is a 32-bit network layer address that comprises a network ID and a host ID. A subnet mask is used to distinguish the network portion of an IP address from the host portion. The 32 bits in an IP address are expressed in dotted decimal notation in four octets (four 8-bit numbers separated by periods, that is, 208.100.50.90). The subnet mask 255.255.255.192 is said to use 26 bits, because it takes 26 of 32 bit positions to create. 255.255.255.192 can be expressed in binary as 11111111.11111111.11111111.11000000, which is three octets using 8 bits each (24) and 2 bits used in the fourth octet to make the number 192. Because the subnet mask is using 26 bits and because an IP address is a total of 32 bits, 6 bits remain to create host IDs on this network. Since the total number of bits used by the subnet mask 255.255.255.192 is 26, all of the other answers are incorrect.

For more information, see Chapter 8 of *CCNA Routing and Switching Exam Cram, Second Edition.*

Question 15

The correct answer is a. If a space is included between a partial or complete command and a question mark, help will interpret this as a keyword request and could result in an "ambiguous command" message. Typing "CL ?" (CL, a space, then a question mark) will result in the message "Ambiguous command: "cl " ", because two commands begin with "CL:" (**CLOCK** and **CLEAR**). Answers b, c, and d are incorrect, because they are invalid Cisco IOS commands. Typing a partial command followed immediately by a question mark (no space between partial command and question mark) will display a list of commands that begin with the partial command entry. Typing "CL?" at the router prompt will return **CLEAR** and **CLOCK**, the two IOS commands that begin with "CL".

For more information, see Chapter 8 of *CCNA Routing and Switching Exam Cram, Second Edition.*

Question 16

The correct answer is g. Pressing the Escape key along with the letter *F* will move the cursor forward one word. None of the other answers will perform this operation, and their correct function is listed below.

Configuring a Cisco router can sometimes involve long, detailed command lines that can be slow to navigate. For this reason, the Cisco IOS includes a number of shortcuts for navigating the command line. The following list details some of the more common methods.

➤ *Ctrl+A*—Pressing the Control key along with the letter *A* will move the cursor to the beginning of the command line.

➤ *Ctrl+E*—Pressing the Control key along with the letter *E* will move the cursor to the end of the command line.

➤ *Ctrl+P*—Pressing the Control key along with the letter *P* will recall previous commands in the command history. (The up arrow key will do this as well.)

➤ *Ctrl+N*—Pressing the Control key along with the letter *N* will return to more recent commands in the command history after recalling previous commands. (The down arrow key will do this as well.)

➤ *Esc+B*—Pressing the Escape key along with the letter *B* will move the cursor back one word.

➤ *Esc+F*—Pressing the Escape key along with the letter *F* will move the cursor forward one word.

➤ *Left and Right Arrow Keys*—These arrow keys will move the cursor one character left and right, respectively.

➤ *Tab*—Pressing the Tab key will complete an entry typed at the router prompt.

For more information, see Chapter 8 of *CCNA Routing and Switching Exam Cram, Second Edition.*

Question 17

The correct answer is d. Typing **TERMINAL NO EDITING** will disable the command history feature on a Cisco router; it is enabled by default. All of the other answers are invalid Cisco IOS commands, therefore, answers a, b, and c are incorrect. The Cisco IOS includes a command history feature that can be used to store previously entered commands in a history buffer.

Commands in the history buffer can be recalled by an operator at the router prompt using Ctrl+P and Ctrl+N, saving configuration time. The command history feature is enabled by default, but it can be disabled, and the size of the history buffer can be modified. The following list details the common commands used when modifying the command history:

➤ *SHOW HISTORY*—Typing the **SHOW HISTORY** command will display all of the commands currently stored in the history buffer. Pressing Ctrl+P will recall previous commands in the command history at the router prompt. (The up arrow key will do this as well.) Pressing Ctrl+N will return to more recent commands in the command history after recalling previous commands. (The down arrow key will do this as well.)

➤ *TERMINAL HISTORY SIZE (0-256)*—Typing **TERMINAL HISTORY SIZE (0-256)** will allow you to set the number of command lines the command history buffer will hold. You must enter a number between 0 and 256. (The default is 10 commands.)

➤ *TERMINAL NO EDITING*—Typing **TERMINAL NO EDITING** will disable the command history feature; it is enabled by default.

➤ *TERMINAL EDITING*—Typing **TERMINAL EDITING** will enable the command history feature if it has been disabled.

For more information, see Chapter 8 of *CCNA Routing and Switching Exam Cram, Second Edition.*

Question 18

The correct answer is a. **COPY TFTP RUNNING-CONFIG** will copy a configuration file from a TFTP server to the router's running configuration in RAM. Answer b is incorrect, because the correct syntax for the **COPY TFTP RUNNING-CONFIG** command specifies a hyphen between **RUNNING** and **CONFIG**. Answers c and d are incorrect, because **COPY TFTP TO RUNNING-CONFIG** and **COPY TFTP SERVER RUNNING-CONFIG** are not valid Cisco IOS commands. The following list outlines the common commands used when working with configuration files:

➤ *COPY STARTUP-CONFIG RUNNING-CONFIG*—This command will copy the startup configuration file in NVRAM to the running configuration in RAM.

➤ *COPY STARTUP-CONFIG TFT*—This command will copy the startup configuration file in NVRAM to a remote TFTP server.

➤ *COPY RUNNING-CONFIG STARTUP-CONFIG*—This command will copy the running configuration from RAM to the startup configuration file in NVRAM.

➤ *COPY RUNNING-CONFIG TFTP*—This command will copy the running configuration from RAM to a remote TFTP server.

➤ *COPY TFTP RUNNING-CONFIG*—This command will copy a configuration file from a TFTP server to the router's running configuration in RAM.

➤ *COPY TFTP STARTUP-CONFIG*—This command will copy a configuration file from a TFTP server to the startup configuration in NVRAM.

Note that the basic syntax of the **COPY** command specifies the source first, then the destination.

To modify a router's running configuration, you must be in global configuration mode. The **CONFIG T** command is used in privileged mode to access global configuration mode. Global configuration mode is displayed at the router prompt with "**config**" in parentheses **(RouterA(config)#)**. To exit global configuration mode, use the keys Control and Z (Ctrl+Z).

For more information, see Chapter 8 of *CCNA Routing and Switching Exam Cram, Second Edition.*

Question 19

The correct answers are d and f. Configuring a Cisco router can sometimes involve long, detailed command lines that can be slow to navigate. For this reason, the Cisco IOS includes a number of shortcuts for navigating the command line. The following list details some of the more common methods:

➤ *Ctrl+A*—Pressing the Control key along with the letter *A* will move the cursor to the beginning of the command line.

➤ *Ctrl+E*—Pressing the Control key along with the letter *E* will move the cursor to the end of the command line.

➤ *Ctrl+P*—Pressing the Control key along with the letter *P* will recall previous commands in the command history. (The up arrow key will do this as well.)

➤ *Ctrl+N*—Pressing the Control key along with the letter *N* will return to more recent commands in the command history after recalling previous commands. (The down arrow key will do this as well.)

➤ *Esc+B*—Pressing the Escape key along with the letter *B* will move the cursor back one word.

➤ *Esc+F*—Pressing the Escape key along with the letter *F* will move the cursor forward one word.

➤ *Left and Right Arrow Keys*—These arrow keys will move the cursor one character left and right, respectively.

➤ Tab—Pressing the Tab key will complete an entry typed at the router prompt.

For more information, see Chapter 8 of *CCNA Routing and Switching Exam Cram, Second Edition.*

Question 20

The correct answers are e, f, and g. The OSI model defines standards used in networking, and comprises a seven-layer model. The presentation layer functions as a translator and is responsible for text and data syntax translation. Presentation layer protocols include PICT (Apple computer picture format), TIFF, JPEG, MIDI, MPEG, and Quick Time. Answers a, b, c, and d are incorrect, because all are application layer protocols. The application layer of the OSI model functions as the "window" for applications to access network resources and is responsible for verifying the identity of the destination node. Application layer protocols include Telnet, FTP, SMTP, WWW, EDI, and WAIS.

For more information, see Chapter 4 of *CCNA Routing and Switching Exam Cram, Second Edition.*

Question 21

The correct answer is a. The **CONFIG T** or **CONFIGURE TERMINAL** command is used in privileged mode to access global configuration mode. Answers b and c are incorrect, because **CONFIG-T** and **CONFIGURE GLOBAL** are invalid Cisco IOS commands. Answer d is incorrect, because the router is in user mode, which displays basic router system information and connects to remote devices. In user mode, the router prompt is followed by an angle bracket **(ROUTER>)**.

Global configuration mode is displayed at the router prompt with "**config**" in parentheses **(RouterA(config)#)** and is used to modify the router's running configuration. To exit global configuration mode, use the keys Control and Z

(Ctrl+Z). Privileged mode is used to modify and view the configuration of the router, as well as to set IOS parameters. Privileged mode also contains the **CONFIGURE** command, which is used to access other configuration modes such as global configuration mode and interface mode. In privileged mode, the router prompt is followed by a pound sign **(ROUTER#)**. To access privileged mode, enter the **ENABLE** command from user mode.

For more information, see Chapter 8 of *CCNA Routing and Switching Exam Cram, Second Edition*.

Question 22

The correct answer is d. The enable secret password is used to control access to privileged mode and will take precedence over the enable password if both are enabled. Answer a is incorrect, because the enable secret password is used to control access to privileged mode, not to control access to the Ethernet port of a router. Answer b is incorrect, because the enable secret password is used to control access to a router even if the enable password is used. Answer c is incorrect, because the auxiliary password is used to control access to any auxiliary ports the router may have. Answer e is incorrect, because the enable secret password is used to control access to privileged mode, not to the console port of a router. Answer f is incorrect, because the enable secret password is used to control access to privileged mode, not user mode. Cisco routers utilize passwords to provide security access to a router. Passwords can be set for controlling access to privileged mode, for controlling access to remote sessions via Telnet, or for controlling access through the auxiliary port. The following list details the common passwords used in Cisco routers:

➤ *Enable Password*—The enable password is used to control access to the privileged mode of the router. If an enable password is specified, it must be entered after the **ENABLE** command to successfully access privileged mode. To configure the enable password, use the **ENABLE PASSWORD** [*password*] command.

➤ *Enable Secret Password*—The enable secret password is used to control access to privileged mode, similar to the enable password. The difference between enable secret and enable password is that the enable secret password will be encrypted for additional security. The enable secret password will take precedence over the enable password if both are enabled. To configure the enable secret password, use the **ENABLE SECRET** [*password*] command.

➤ *Virtual Terminal Password*—The virtual terminal password is used to control remote Telnet access to a router. Setting the virtual terminal password will require all users that Telnet into the router to provide this password for access. To configure the virtual terminal password, use the following commands:

```
LINE VTY 0 4
   LOGIN
   PASSWORD [password]
```

➤ *Auxiliary Password*—The auxiliary password is used to control access to any auxiliary ports the router may have. Setting the auxiliary password will require all users connecting to the auxiliary port of the router (usually remote dial-in) to provide this password for access. To configure the auxiliary password, use the following commands:

```
LINE AUX 0
   LOGIN
   PASSWORD [password]
```

➤ *Console Password*—The console password is used to control access to the console port of the router. Setting the console password will require all users connecting to the router console to provide this password for access. To configure the console password, use the following commands:

```
LINE CON 0
   LOGIN
   PASSWORD [password]
```

For more information, see Chapter 8 of *CCNA Routing and Switching Exam Cram, Second Edition.*

Question 23

The correct answer is d. The **BANNER MOTD** command must be entered in global configuration mode, which can be accessed from privileged mode using the **CONFIG T** command. When a router is in global configuration mode, the prompt displays "**config**" in parentheses (**Router(config)#**). Answer a is incorrect, because you can use a pound (#) sign as a delimiting character. Answers b and c are incorrect, because the **BANNER MOTD** command cannot be entered in user mode. A Cisco router can be configured to display a message when users log in; this message is known as a logon banner and can be configured using the **BANNER MOTD** command. The correct syntax for setting a logon banner is as follows:

```
BANNER MOTD # (Note that the pound sign [#] is a delimiting charac-
ter of your       choice.)
   [Message text]
   #
```

An example of a logon banner would be:

```
Router(config)#banner motd #
   Enter TEXT message. End with the character '#'.
   This is the message text of my log on banner #
   Router(config)#^Z
```

For more information, see Chapter 8 of *CCNA Routing and Switching Exam Cram, Second Edition.*

Question 24

The correct answer is c. Because the subnet mask 255.255.224.0 is using 19 bits and because an IP address is a total of 32 bits, 13 bits remain to create host IDs on this network. An IP address is a 32-bit network layer address that comprises a network ID and a host ID. A subnet mask is used to distinguish the network portion of an IP address from the host portion. The 32 bits in an IP address are expressed in dotted decimal notation in 4 octets (four 8-bit numbers separated by periods, that is, 208.100.50.90). The subnet mask 255.255.224.0 is said to use 19 bits, because it takes 19 of 32 bit positions to create. 255.255.224.0 can be expressed in binary as 11111111. 11111111.11100000.00000000, which is two octets using 8 bits each (16) and 3 bits used in the third octet to make the number 224. Since the number of bits to create hosts is 13, all of the other answers are incorrect.

For more information, see Chapter 13 of *CCNA Routing and Switching Exam Cram, Second Edition.*

Question 25

The correct answer is d. **ROUTER(config-if)#IP ACCESS-GROUP 88 OUT** is the correct command to use in this question to apply the access list 88 to the outgoing interface of the router. For an IP access list to filter traffic, it must be applied to an interface. This is done using the **IP ACCESS-GROUP** command. This command must be entered in interface configuration mode. To access this mode, use the **INTERFACE** *[interface-number]* command from global configuration mode. When the router is in interface configuration mode, the router prompt displays "**config-if**" in parentheses, followed by a pound sign (**(ROUTER(config-if)#)**. Answers a and b are incorrect, because the router is in privileged mode. Answer c is incorrect, because the correct syntax for the **IP ACCESS-GROUP** command begins with **IP**. Answer e is incorrect, because the correct syntax for the **IP ACCESS GROUP** command contains a hyphen between **ACCESS** and **GROUP**. Answer f is incorrect, because the router is in user mode. The **IP ACCESS-GROUP** command has the following syntax:

```
IP ACCESS-GROUP [access-list number] [out|in]
```

The **[out|in]** parameter specifies where the access list will be applied on the router. **OUT** will cause the router to apply the access list to all outgoing packets, and **IN** will cause the router to apply the access list to all incoming packets.

Access lists can be used to filter the traffic that is handled by a Cisco router. An IP access list will filter IP traffic and can be created as a standard IP access list or an extended IP access list. A standard IP access list can be used to permit or deny access based on IP addressing information only and can only act on the source IP addressing information.

For more information, see Chapter 4 of *CCNA Routing and Switching Exam Cram, Second Edition.*

Question 26

The correct answer is b. The **IP ACCESS-GROUP 155 OUT** is the correct command to use in this question to apply the access list **155** to the outgoing interface of the router. Answer a is incorrect, because the wrong access list number, 10, is specified. Answer c is incorrect, because the **IP ACCESS-GROUP** begins with **IP**. Answer d is incorrect, because the interface number, **S0**, is specified instead of the access list number, **155**. For an IP access list to filter traffic, it must be applied to an interface. This is done using the **IP AC-CESS-GROUP** command. This command must be entered in interface configuration mode. To access this mode, use the **INTERFACE** *[interface-number]* command from global configuration mode. When the router is in interface configuration mode, the router prompt displays "**config-if**" in parentheses, followed by a pound sign (**ROUTER(config-if)#**). The **IP ACCESS GROUP** command has the following syntax:

```
IP ACCESS-GROUP [access-list number] [out|in]
```

The [out/in] parameter specifies where the access list will be applied on the router. OUT will cause the router to apply the access list to all outgoing packets, and IN will cause the router to apply the access list to all incoming packets.

For more information, see Chapter 13 of *CCNA Routing and Switching Exam Cram, Second Edition.*

Question 27

The correct answer is d. The OSI model defines standards used in networking and comprises a seven-layer model. The transport layer of the OSI model is responsible for flow control, multiplexing, and end-to-end communication. Since no other layer of the OSI model is responsible for end-to-end communication, all of the other answers are incorrect.

The following list outlines the seven layers of the OSI model and their functions:

➤ *Application*—The "window" to networking used by programs, the application layer is responsible for:

  ➤ Verifying that the appropriate resources are present to initiate a connection with the destination node.

  ➤ Verifying the identity of the destination node.

Application layer protocols include Telnet, FTP, SMTP, WWW, EDI, and WAIS.

➤ *Presentation*—Essentially a translator, the presentation layer is responsible for:

  ➤ Translating text and data syntax, such as EBCDIC, used in IBM systems, to ASCII, used in PCs and most computer systems.

  ➤ Using Abstract Syntax Notation One (ASN 1) to perform data translation.

esentation layer protocols include PICT, TIFF, JPEG, MIDI, MPEG, and Quick Time (audio/video application).

➤ *Session*—Used to coordinate communication between nodes, the session layer is responsible for:

  ➤ Creating, maintaining, and ending a communication session.

  ➤ Coordinating service requests and responses that occur between nodes across the network.

Session layer protocols include Network File System (NFS), used by SUN Microsystems and Unix with TCP/IP; structured query language (SQL, used to define database information requests), Remote Procedure Calls (RPCs, used in Microsoft network communication), X Window (used by Unix terminals), AppleTalk Session Protocol (ASP, used by Apple computers), and Digital Network Architecture Session Control Protocol (DNA SCP, used by IBM).

➤ *Transport*—Reliable end-to-end communication is the primary function of the transport layer. Its many responsibilities include:

  ➤ Ensuring flow control so the amount of data being sent from one node will not overwhelm the destination node.

  ➤ Ensuring the ability of multiple applications to utilize a single transport (multiplexing).

  ➤ Ensuring the reliable transfer of data. The transport layer implements connection-oriented services between nodes, which utilize a three-way handshake (synchronization, acknowledgment, and data-transfer) to efficiently transfer data.

  ➤ Ensuring positive acknowledgment, or the process of a node waiting for an acknowledgment from the destination node prior to sending data.

  ➤ Windowing, a form of flow control that specifies how much data will be transferred between acknowledgments.

Transport layer protocols include TCP, part of the TCP/IP protocol suite, and Sequenced Packet Exchange (SPX), part of the IPX/SPX protocol suite.

➤ *Network*—Selecting the appropriate path a packet should take to get to its intended destination is the function of the network layer. It is responsible for routing, which is the process of using a network layer address to determine the best path a packet will travel to its destination. Network layer protocols include IP and IPX, part of the IPX/SPX protocol suite.

➤ *Data Link*—Divided into two sublayers Media Access Control (MAC) and Logical Link Control (LLC), the data-link layer is responsible for:

  ➤ Preparing data from upper layers to be transmitted over the physical medium by encapsulating upper layer data into frames. This frame includes the source and destination of MAC addresses in the frame header.

  ➤ Converting data into bits, so it can be transmitted by the physical layer.

> ➤ Adding a cyclical redundancy check (CRC) to the end of a frame, which is used for error checking at the data-link layer.

Data-link protocols include High-Level Data Link Control (HDLC, the Cisco default encapsulation for serial connections), Synchronous Data Link Control (SDLC, used in IBM networks), Link Access Procedure Balanced (LAPB, used with X.25), X.25 (a packet switching network), Serial Line Internet Protocol (SLIP, an older TCP/IP dial-up protocol), Point-to-Point Protocol (PPP, a newer dial-up protocol), ISDN, a dial-up digital service; and frame relay, a packet switching network.

> ➤ *Physical*—The lowest layer of the OSI model, the physical layer defines the electrical functionality required to send and receive bits over a given physical medium. Specifications that define the voltage levels and physical components of a network are defined at the physical layer. Protocols are not specified at the physical layer, because they are implemented as software. Examples of the standards for sending data over the physical medium are Ethernet (the most widely used standard in networking), Token Ring (IBM's proprietary networking topology), and Fiber Distributed Data Interface (FDDI, a standard for fiber optic networks, commonly used as a backbone).

For more information, see Chapter 4 of *CCNA Routing and Switching Exam Cram, Second Edition.*

Question 28

The correct answer is a. A "-1" used in the source or destination network of a standard IPX access list means "all networks." Answers b, c, d, e, f, and g are incorrect. Access lists can be used to filter the traffic that is handled by a Cisco router. An IPX access list will filter IPX traffic and can be created as a standard IPX access list or an extended IPX access list. A standard IPX access list can be used to permit or deny access based on the source and destination IPX address information. The syntax for creating a standard IPX access list is:

```
ACCESS-LIST [access-list-number] [permit|deny] [protocol] source
    [source-node-mask] [socket][destination]
    [[destination-node-mask socket]
```

> ➤ *ACCESS-LIST*—This command is followed by the **access list number**. When using an extended IPX access list, this can be a number between 900 and 999.

➤ *[permit\deny]*—This statement will either allow or disallow traffic in this access list.

➤ *[protocol]*—This statement is used to specify an IPX protocol. A "-1" can be used to specify all protocols.

➤ *source*—This statement is used to specify the source IPX addressing information. A "-1" can be used to signify "all networks."

➤ *[socket]*—The socket number of the application. A "0" can be used to specify all sockets.

➤ *[destination]*—This statement is used to specify destination addressing information. (A "-1" may be used here as well.)

➤ *[source-node-mask]* and *[destination-node-mask]*—These are optional parameters for controlling access to a portion of a network, similar to the wildmask that is used in IP access lists.

For more information, see Chapter 13 of *CCNA Routing and Switching Exam Cram, Second Edition.*

Question 29

The correct answer is c. The **SHOW IP ACCESS-LISTS** command will display a list of all the IP access lists that are currently in use on your router, as well as what lines in those access lists were matched by the traffic that passed through the router. Answer a is incorrect, because no hyphen is between **ACCESS** and **LISTS**. Answers b and d are incorrect, because **SHOW IP INTERFACE** will not display any matches to the individual lines of the access list, as the **SHOW ACCESS-LIST** commands do. Access lists can be used to filter the traffic handled by a Cisco router. Several Cisco IOS commands can be used to view the various access lists in use on a Cisco router. The following list details the most common Cisco IOS commands used to view access lists:

➤ *SHOW ACCESS-LISTS*—This command will display all of the access lists in use on the router. It will also show each line of the access list and return the number of times a packet matched that line. This command can also be used with a specific access list number to display this detail on a single IP access list. **SHOW ACCESS-LISTS** will not display what interface an access list has been applied to, as the **SHOW INTERFACE** commands do.

➤ *SHOW IP ACCESS-LIST*—This command will display all IP access lists in use on the router. It will also show each line of the access list and return the number of times a packet matched that line. This command can also be used with a specific IP access list number to display this detail on a single IP access list. **SHOW IP ACCESS-LISTS** will not display what interface an access list has been applied to, as the **SHOW INTERFACE** commands do.

➤ *SHOW IPX ACCESS-LIST*—This command will display all IP access lists in use on the router. It will also show each line of the access list and return the number of times a packet matched that line. This command can also be used with a specific IPX access list number to display this detail on a single IPX access list. **SHOW IPX ACCESS-LISTS** will not display what interface an access list has been applied to, as the **SHOW INTERFACE** commands do.

➤ *SHOW IP INTERFACE [interface number]*—This command will display the IP access lists that have been applied to a specific interface. It will not display any matches to the individual lines of the access list, as the **SHOW ACCESS-LIST** commands do.

➤ *SHOW IPX INTERFACE [interface number]*—This command will display the IP access lists that have been applied to a specific interface. It will not display any matches to the individual lines of the access list, as the **SHOW ACCESS-LIST** commands do.

For more information, see Chapter 13 of *CCNA Routing and Switching Exam Cram, Second Edition.*

Question 30

The correct answers are a, b, d, e, and g. The OSI model defines standards used in networking and comprises a seven-layer model. The data-link layer of the OSI model is responsible for preparing data from upper layers to be transmitted over the physical medium by encapsulating upper layer data into frames. Data-link protocols include HDLC, the Cisco default encapsulation for serial connections; SDLC, used in IBM networks; LAPB, used with X.25; X.25, a packet switching network; SLIP, an older TCP/IP dial-up protocol; PPP, a newer dial-up protocol; ISDN, a dial-up digital service; and frame relay, a packet switching network. Answers c and f are incorrect, because RPCs and NFS (used by SUN Microsystems and Unix with TCP/IP) are session layer protocols.

For more information, see Chapter 4 of *CCNA Routing and Switching Exam Cram, Second Edition.*

Question 31

The correct answer is a. The **SHOW IPX INTERFACE** command will display the interface that IPX SAP access list 1091 has been applied to. Answer b is incorrect, because a hyphen is between **IPX** and **INTERFACE**. Answers c and d are incorrect, because **SHOW SAP INTERFACE** and **SHOW SAP-INTERFACE** are not valid Cisco IOS commands. Access lists can be used to filter the traffic that is handled by a Cisco router. Several Cisco IOS commands can be used to view the various access lists in use on a Cisco router. The following list details the most common Cisco IOS commands used to view access lists:

> *SHOW ACCESS-LISTS*—This command will display all of the access lists in use on the router. It will also show each line of the access list and return the number of times a packet matched that line. This command can also be used with a specific access list number to display this detail on a single IP access list. **SHOW ACCESS-LISTS** will not display what interface an access list has been applied to, as the **SHOW INTERFACE** commands do.

> *SHOW IP ACCESS-LIST*—This command will display all IP access lists in use on the router. It will also show each line of the access list and return the number of times a packet matched that line. This command can also be used with a specific IP access list number to display this detail on a single IP access list. **SHOW IP ACCESS-LISTS** will not display what interface an access list has been applied to, as the **SHOW INTERFACE** commands do.

> *SHOW IPX ACCESS-LIST*—This command will display all IP access lists in use on the router. It will also show each line of the access list and return the number of times a packet matched that line. This command can also be used with a specific IPX access list number to display this detail on a single IPX access list. **SHOW IPX ACCESS-LISTS** will not display what interface an access list has been applied to, as the **SHOW INTERFACE** commands do.

> *SHOW IP INTERFACE [interface number]*—This command will display the IP access lists that have been applied to a specific interface. It will not display any matches to the individual lines of the access list, as the **SHOW ACCESS-LIST** commands do.

> *SHOW IPX INTERFACE [interface number]*—This command will display the IP access lists that have been applied to a specific interface. It will not display any matches to the individual lines of the access list, as the **SHOW ACCESS-LIST** commands do.

For more information, see Chapter 12 of *CCNA Routing and Switching Exam Cram, Second Edition.*

Question 32

The correct answer is a. Customer premises equipment (CPE) are the communication devices (telephones, modems, and so forth) that exist at the customer's location. Answer b is incorrect, because the CO, or Central Office, is the telephone company's premises that act as the central communication point for the customer. Answer c is incorrect, because the local loop is the cabling that extends from the DEMARC to the CO. Answer d is incorrect, because answer a is correct. The most common form of WAN connectivity utilizes telecommunication technology. When WANs communicate over telecommunication lines, several components are used to facilitate the connection. The basic process of WAN communication begins with a call placed using a CPE that connects to the DEMARC, which passes the data through the local loop to the central office (CO). The CO will act as the central distribution point for sending and receiving information. The following list details the common components of WAN communication:

➤ *Customer Premises Equipment (CPE)*—CPE devices are the communication devices (telephones, modems, and so forth) that exist at the customer's location.

➤ *Demarcation (DEMARC)*—Short for demarcation, the DEMARC is the point at the customer's premises where CPE devices connect. The telephone company owns the DEMARC, which is usually a large punch-down board located in a wiring closet.

➤ *Local Loop*—The local loop is the cabling that extends from the DEMARC to the telephone company's CO.

➤ *Central Office (CO)*—The central office is the telephone company's office that acts as the central communication point for the customer.

For more information, see Chapter 11 of *CCNA Routing and Switching Exam Cram, Second Edition.*

Question 33

The correct answer is b. Cisco routers support three Local Management Interface (LMI) signaling types: CISCO, ANSI, and Q933A. Answer a is incorrect, because Cisco supports three LMI signalling types. Answer c is incorrect, because the default signal type is Cisco. Answer d is incorrect, because the default LMI signal type is Cisco. The Cisco default LMI type is CISCO. Frame relay is a packet switching technology that utilizes logical circuits to form a connection. This logical connection is identified by the DLCI (data-link connection identifier). The DLCI is a unique identifier used to identify a specific frame relay connection, which is managed through a signaling format known as LMI. The LMI interface provides information about the DLCI values, as well as other connection-related information. Your Cisco router must utilize the same LMI signaling format as your service provider. There are three LMI signaling formats: CISCO, ANSI, and Q933A. The Cisco default LMI type is CISCO.

For more information, see Chapter 11 of *CCNA Routing and Switching Exam Cram, Second Edition.*

Question 34

The correct answer is b. Protocols that begin with the letter E (such as E.164) define ISDN standards on the existing telephone network. Answer a is incorrect, because protocols that begin with the letter I (such as I.430 and I.431) specify ISDN concepts, interfaces, and terminology. Answer d is incorrect, because protocols that begin with the letter Q specify ISDN switching and signaling standards.

ISDN is a communications standard that uses digital telephone lines to transmit voice, data, and video. ISDN is a dial-up service that is used on demand, and its protocol standards are defined by the International Telecommunication Union (ITU). Protocols that begin with the letter E (such as E.164) define ISDN standards on the existing telephone network. Protocols that begin with the letter Q (such as Q.931 and Q.921) specify ISDN switching and signaling standards, and protocols that begin with the letter I (such as I.430 and I.431) specify ISDN concepts, interfaces, and terminology. The bandwidth provided by ISDN can be divided into two categories: Primary Rate ISDN (PRI) and Basic Rate ISDN (BRI). Basic Rate ISDN (also known as 2B+1D) comprises two B channels that are 64Kbps each and one 16Kbps D channel. Basic Rate ISDN uses the two B channels to transmit data and uses the 16Kbps D channel for control and signaling purposes. Primary Rate ISDN (also known as 23B+1D) comprises 23 B channels that are 64KBps each and one that is a 64Kbps D channel.

For more information, see Chapter 11 of *CCNA Routing and Switching Exam Cram, Second Edition.*

Question 35

The correct answer is a. Typing "**TERMINAL HISTORY SIZE (0-256)**" will allow you to set the number of command lines the command history buffer will hold. All of the other answers are invalid Cisco IOS commands, therefore, answers b, c, and d are incorrect. The Cisco IOS includes a command history feature that can be used to store previously entered commands in a history buffer. Commands in the history buffer can be recalled by an operator at the router prompt, using Ctrl+P and Ctrl+N, saving configuration time. The command history feature is enabled by default, but it can be disabled, and the size of the history buffer can be modified. The following list details the common commands used when modifying the command history:

➤ *SHOW HISTORY*—Typing the **SHOW HISTORY** command will display all of the commands currently stored in the history buffer. Pressing Ctrl+P will recall previous commands in the command history at the router prompt. (The up arrow key will do this as well.) Pressing Ctrl+N will return to more recent commands in the command history after recalling previous commands. (The down arrow key will do this as well.)

➤ *TERMINAL HISTORY SIZE (0-256)*—Typing the **TERMINAL HISTORY SIZE (0-256)** command will allow you to set the number of command lines the command history buffer will hold. You must enter a number between 0 and 256. (The default is 10 commands.)

➤ *TERMINAL NO EDITING*—Typing the **TERMINAL NO EDITING** command will disable the command history feature. It is enabled by default.

➤ *TERMINAL EDITING*—Typing the **TERMINAL EDITING** command will enable the command history feature if it has been disabled.

For more information, see Chapter 8 of *CCNA Routing and Switching Exam Cram, Second Edition*.

Question 36

The correct answer is b. The node ID of this IPX address is 1111.2222.3333. Answers a, c, and d are incorrect, because they are erroneous combinations of numbers. The IPX/SPX protocol is used in Novell NetWare networks. An IPX address is a network layer, hierarchical address that is similar in format to an IP address in that IPX addresses contain a network ID and a node ID. The network ID of an IPX address identifies the network that an IPX node is a part of and will be the same for all IPX computers in the same network. The node ID of an IPX address is unique to each host on an IPX network, and is assigned dynamically using the MAC address of the computer.

The network ID of an IPX address can be up to 32 bits in length. The network ID is usually expressed in decimal notation without the leading zeros. The node ID of an IPX address is 48 bits in length and can be expressed in decimal notation by four digits separated by three periods. A complete IPX address is expressed in decimal notation by combining the network ID with the node ID. For example, if an IPX host has a node ID of 1111.2222.3333 and is part of IPX network 5F, the IPX address for that host would be 5F.1111.2222.3333.

For more information, see Chapter 12 of *CCNA Routing and Switching Exam Cram, Second Edition.*

Question 37

The correct answer is b. The **CONFIGURE OVERWRITE-NETWORK (CONFIG O)** command can be abbreviated as **CONFIG O** and must be entered from privileged mode. In privileged mode, the router prompt is followed by a pound sign **(ROUTER#)**. Answer a is incorrect, because the router is in user mode. The **CONFIG O** command cannot be entered from user mode. In user mode, the router prompt is followed by an angle bracket **(ROUTER>)**. Answer c is incorrect, because the prompt is already in configuration mode. After entering configuration mode, the router prompt displays "**config**" in parentheses **(ROUTER(config)#)**. Answer d is incorrect, because the prompt is displaying interface mode **(ROUTER(CONFIG-IF)#)**.

For more information, see Chapter 8 of *CCNA Routing and Switching Exam Cram, Second Edition.*

Question 38

The correct answer is a. Flow control is a method used by TCP at the transport layer that ensures the amount of data being sent from one node will not overwhelm the destination node. Three basic techniques are used in flow control:

➤ *Windowing*—This is the process of predetermining the amount of data that will be sent before an acknowledgment is expected.

➤ *Buffering*—This is a method of temporarily storing packets in memory buffers until they can be processed by the computer.

➤ *Source-Quench Messaging*—This is a message sent by a receiving computer when its memory buffers are nearing capacity. A source-quench message is a device's way of telling the sending computer to slow down the rate of transmission. Multiplexing is the ability of multiple applications to utilize a single transport and is not used as part of flow control.

Answers b and c are incorrect, because multiplexing is not used in flow control. Answer d is incorrect, because answer a correctly states the methods used in flow control.

For more information, see Chapter 9 of *CCNA Routing and Switching Exam Cram, Second Edition.*

Question 39

The correct answers are d, e, and f. Answer a is incorrect, because the *TOKEN RING_SNAP* frame used by Novell is called SNAP in Cisco. Answer b is incorrect, because the ETHERNET 802.2 frame used by Novell is called SAP in Cisco. Answer c is incorrect, because the ETHERNET_snap frame used in Novell is called SNAP in Cisco.

The IPX/SPX protocol is used in Novell NetWare networks. When IPX packets are passed to the data-link layer, they are encapsulated into frames for transmission over the physical media. Novell NetWare supports a number of different encapsulation methods or frame types used to encapsulate IPX packets. The frame type used by NetWare computers must be the same as the one used by a Cisco router, or they will not be able to communicate. When configuring a Cisco router to support IPX, you can specify the frame type to use. Novell NetWare and Cisco both use different names to describe the same frame type. The following list details the frame types used by NetWare networks and the comparable Cisco name.

➤ *ETHERNET II*—This is used by Novell and is called ARPA in Cisco.

➤ *ETHERNET 802.2*—This is used by Novell and is called SAP in Cisco.

➤ *ETHERNET_snap*—This is used by Novell and is called SNAP in Cisco.

➤ *ETHERNET 802.3*—This is used by Novell and is called NOVELL-ETHER in Cisco and is the default frame type.

➤ *TOKEN RING*—This is used by Novell and is called TOKEN in Cisco.

➤ *TOKEN RING_snap*—This is used by Novell and is called SNAP in Cisco.

For more information, see Chapter 12 of *CCNA Routing and Switching Exam Cram, Second Edition.*

Question 40

The correct answer is b. Open Shortest Path First (OSPF) is considered a link state routing protocol. Answer a is incorrect, because RIP is a distance vector routing protocol. Answer c is incorrect, because Enhanced Interior Gateway Routing Protocol (EIGRP) is considered a balanced hybrid routing protocol that combines the features of link state and distance vector. Answer d is incorrect, because IGRP is a distance vector routing protocol. A Cisco router cannot route anything unless an entry is in its routing table that contains information on where to send the packet.

A routing table can be of two basic types: static and dynamic. A static routing table is one that has its entries entered manually by an operator. On a Cisco router, a static route can be entered using the **IP ROUTE** command. A dynamic routing table is one that has its entries entered automatically by the router through the use of a routing protocol. A dynamic routing protocol provides for automatic discovery of routes and eliminates the need for static routes. Routing protocols generally fall into one of two categories: link state and distance vector. Distance vector routing protocols dynamically update their routing tables by broadcasting their own routing information at specified intervals, and they include RIP and IGRP. Link state routing protocols, such as OSPF, update their routing tables by sending multicast "hello" messages to their neighbors and only send updated information when a change occurs in the network.

For more information, see Chapter 11 of *CCNA Routing and Switching Exam Cram, Second Edition*.

Question 41

The correct answer is e. The OSI model defines standards used in networking and comprises a seven-layer model. The network layer of the OSI model is responsible for routing and makes use of logical addresses to determine the path a packet will take to get to its destination. Because no other layer of the OSI model is responsible for routing, all of the other answers are incorrect.

The following list outlines the seven layers of the OSI model and their functions:

➤ *Application*—The "window" to networking used by programs, the application layer is responsible for:

➤ Verifying that the appropriate resources are present to initiate a connection with the destination node.

➤ Verifying the identity of the destination node.

Application layer protocols include Telnet, FTP, SMTP, WWW, EDI, and WAIS.

➤ *Presentation*—Essentially a translator, the presentation layer is responsible for:

   ➤ Translating text and data syntax, such as EBCDIC, used in IBM systems, to ASCII, used in PCs and most computer systems.

   ➤ Using ASN 1 to perform data translation.

Presentation layer protocols include PICT, TIFF, JPEG, MIDI, MPEG, and Quick Time.

➤ *Session*—Used to coordinate communication between nodes, the session layer is responsible for:

   ➤ Creating, maintaining, and ending a communication session.

   ➤ Coordinating service requests and responses that occur between nodes across the network.

Session layer protocols include NFS, used by SUN Microsystems and Unix with TCP/IP; SQL, used to define database information requests; RPC, used in Microsoft network communication; X Window, used by Unix terminals; ASP, used by Apple computers; and DNA SCP, used by IBM.

➤ *Transport*—Reliable end-to-end communication is the primary function of the transport layer. Its many responsibilities include:

   ➤ Ensuring flow control, so the amount of data being sent from one node will not overwhelm the destination node.

   ➤ Ensuring the ability of multiple applications to utilize a single transport (multiplexing).

   ➤ Ensuring the reliable transfer of data. The transport layer implements connection-oriented services between nodes, which utilize a three-way handshake (synchronization, acknowledgment, and data-transfer) to efficiently transfer data.

   ➤ Ensuring positive acknowledgment, or the process of a node waiting for an acknowledgment from the destination node prior to sending data.

   ➤ Windowing, a form of flow control that specifies how much data will be transferred between acknowledgments.

Transport layer protocols include TCP, part of the TCP/IP protocol suite, and SPX, part of the IPX/SPX protocol suite.

➤ *Network*—Selecting the appropriate path a packet should take to get to its intended destination is the function of the network layer. It is responsible for routing, which is the process of using a network layer address to determine the best path a packet will travel to its destination. Network layer protocols include IP and IPX, part of the IPX/SPX protocol suite.

➤ *Data Link*—Divided into two sublayers, MAC and LLC, the data-link layer is responsible for:

  ➤ Preparing data from upper layers to be transmitted over the physical medium by encapsulating upper layer data into frames. This frame includes the source and destination of MAC addresses in the frame header.

  ➤ Converting data into bits, so it can be transmitted by the physical layer.

  ➤ Adding a CRC to the end of a frame, which is used for error checking at the data-link layer.

Data-link protocols include HDLC, the Cisco default encapsulation for serial connections; SDLC, used in IBM networks; LAPB, used with X.25; X.25, a packet switching network; SLIP, an older TCP/IP dial-up protocol; PPP, a newer dial-up protocol; ISDN, a dial-up digital service; and frame relay, a packet switching network.

➤ *Physical*—The lowest layer of the OSI model, the physical layer defines the electrical functionality required to send and receive bits over a given physical medium. Specifications that define the voltage levels and physical components of a network are defined at the physical layer. Protocols are not specified at the physical layer, because they are implemented as software. Examples of the standards for sending data over the physical medium are Ethernet (the most widely used standard in networking), Token Ring (IBM's proprietary networking topology), and FDDI, a standard for fiber optic networks, commonly used as a backbone.

For more information, see Chapter 4 of *CCNA Routing and Switching Exam Cram, Second Edition.*

Question 42

The correct answers are b and d. The Cisco IOS can be permanently stored in a Cisco router in either the Flash or ROM. ROM is a physical chip installed on a router's motherboard that contains the bootstrap program, the POST, and the operating system software (Cisco IOS). Answer a is incorrect, because no information can be permanently stored in RAM. Answer c is incorrect, because NVRAM is used by Cisco routers to store the startup or running configuration, not the IOS.

When a Cisco router is first powered up, the bootstrap program and the POST are executed. ROM cannot be changed through software; the chip must be replaced if any modifications are needed. The Flash is essentially the programmable read-only memory (PROM) used in PCs, which is an erasable, programmable memory area that contains the Cisco operating system software (IOS). When a router's operating system needs to be upgraded, new software can be downloaded from a TFTP server to the router's Flash. Upgrading the IOS in this manner is typically more convenient than replacing the ROM chip in the router's motherboard. Information in Flash is retained when the router is powered off.

For more information, see Chapter 3 of *CCNA Routing and Switching Exam Cram, Second Edition.*

Question 43

The correct answer is c. This access list will allow traffic from all hosts on network 131.200.0.0. Answer a is incorrect, because the access list is using a wildmask, which will allow access to all hosts on the network, not just one. Answer b is incorrect, because a wildmask is not used to specify a network ID. Answer d is incorrect, because this is a valid access list.

Access lists can be used to filter the traffic that is handled by a Cisco router. An IP access list will filter IP traffic and can be created as a standard IP access list or an extended IP access list. A standard IP access list can be used to permit or deny access based on IP addressing information only and can only act on the source IP addressing information. The syntax for creating a standard IP access list is:

```
ACCESS-LIST [access-list-number] [permit|deny] source
    [wildcard mask]
```

➤ *ACCESS-LIST*—This command is followed by the **access list number**. When using a standard IP access list, this can be a number between 1 and 99.

➤ *[permit|deny]*—This statement will either allow or disallow traffic in this access list.

➤ *source*—This statement is used to specify the source IP addressing information. A standard IP access list can only act on a source address.

➤ *[wildcard mask]*—This is an optional parameter in a standard IP access list. If no wildcard mask is specified, the access list will only act on a single IP address that is specified in the **source**. A wildmask can be used to control access to an entire IP network ID rather than a single IP address. A wildcard mask will use the number 255 to mean "any" and the number 0 to mean "must match exactly."

For more information, see Chapter 13 of *CCNA Routing and Switching Exam Cram, Second Edition.*

Question 44

The correct answer is d. The **COPY STARTUP-CONFIG RUNNING-CONFIG** will copy the startup configuration file in NVRAM to the running configuration in RAM. Answer a is incorrect, because the **COPY RUNNING-CONFIG STARTUP CONFIG** command will copy the running configuration from RAM to the startup configuration file in NVRAM. Answers b and c are incorrect, because **WRITE STARTUP-CONFIG RUNNING-CONFIG** and **WRITE RUNNING-CONFIG STARTUP-CONFIG** are not valid Cisco IOS commands.

The following list outlines the common commands used when working with configuration files:

➤ *COPY STARTUP-CONFIG RUNNING-CONFIG*—This command will copy the startup configuration file in NVRAM to the running configuration in RAM.

➤ *COPY STARTUP-CONFIG TFTP*—This command will copy the startup configuration file in NVRAM to a remote TFTP server.

➤ *COPY RUNNING-CONFIG STARTUP-CONFIG*—This command will copy the running configuration from RAM to the startup configuration file in NVRAM.

➤ *COPY RUNNING-CONFIG TFTP*—This command will copy the running configuration from RAM to a remote TFTP server.

➤ *COPY TFTP RUNNING-CONFIG*—This command will copy a configuration file from a TFTP server to the router's running configuration in RAM.

➤ *COPY TFTP STARTUP-CONFIG*—This command will copy a configuration file from a TFTP server to the startup configuration in NVRAM.

Note that the basic syntax of the **COPY** command specifies the source first, then the destination.

To modify a router's running configuration, you must be in global configuration mode. The **CONFIG T** command is used in privileged mode to access global configuration mode. This mode is displayed at the router prompt with "**config**" in parentheses (**RouterA(config)#**). To exit global configuration mode, use the keys Control and Z (Ctrl+Z).

For more information, see Chapter 8 of *CCNA Routing and Switching Exam Cram, Second Edition.*

Question 45

The correct answer is g. The OSI model defines standards used in networking and comprises a seven-layer model. The physical layer of the OSI model defines the voltage levels and physical components of a network. Since no other layer of the OSI model is responsible for defining voltage levels, all of the other answers are incorrect.

The following list outlines the seven layers of the OSI model and their functions:

➤ *Application*—The "window" to networking used by programs, the application layer is responsible for:

   ➤ Verifying that the appropriate resources are present to initiate a connection with the destination node.

   ➤ Verifying the identity of the destination node.

Application layer protocols include Telnet, FTP, SMTP, WWW, EDI, and WAIS.

➤ *Presentation*—Essentially a translator, the presentation layer is responsible for:

   ➤ Translating text and data syntax, such as EBCDIC, used in IBM systems, to ASCII, used in PCs and most computer systems.

   ➤ Using ASN 1 to perform data translation.

Presentation layer protocols include PICT, TIFF, JPEG, MIDI, MPEG, and Quick Time.

➤ *Session*—Used to coordinate communication between nodes, the session layer is responsible for:

➤ Creating, maintaining, and ending a communication session.

➤ Coordinating service requests and responses that occur between nodes across the network.

Session layer protocols include NFS, used by SUN Microsystems and Unix with TCP/IP; SQL, used to define database information requests; RPCs, used in Microsoft network communication; X Window, used by Unix terminals; ASP, used by Apple computers; and DNA SCP, used by IBM.

➤ *Transport*—Reliable end-to-end communication is the primary function of the transport layer. Its many responsibilities include:

➤ Ensuring flow control, so the amount of data being sent from one node will not overwhelm the destination node.

➤ Ensuring the ability of multiple applications to utilize a single transport (multiplexing).

➤ Ensuring the reliable transfer of data. The transport layer implements connection-oriented services between nodes, which utilize a three-way handshake (synchronization, acknowledgment, and data-transfer) to efficiently transfer data.

➤ Ensuring positive acknowledgment, or the process of a node waiting for an acknowledgment from the destination node prior to sending data.

➤ Windowing, a form of flow control that specifies how much data will be transferred between acknowledgments.

Transport layer protocols include TCP, part of the TCP/IP protocol suite, and SPX, part of the IPX/SPX protocol suite.

➤ *Network*—Selecting the appropriate path a packet should take to get to its intended destination is the function of the network layer. It is responsible for routing, which is the process of using a network layer address to determine the best path a packet will travel to its destination. Network layer protocols include IP and IPX, part of the IPX/SPX protocol suite.

➤ *Data Link*—Divided into two sublayers, MAC and LLC, the data-link layer is responsible for:

➤ Preparing data from upper layers to be transmitted over the physical medium by encapsulating upper layer data into frames. This frame includes the source and destination of MAC addresses in the frame header.

➤ Converting data into bits, so it can be transmitted by the physical layer.

➤ Adding a CRC to the end of a frame, which is used for error checking at the data-link layer.

Data-link protocols include HDLC, the Cisco default encapsulation for serial connections; SDLC, used in IBM networks; LAPB, used with X.25; X.25, a packet switching network; SLIP, an older TCP/IP dial-up protocol; PPP, a newer dial-up protocol; ISDN, a dial-up digital service; and frame relay, a packet switching network.

➤ *Physical*—The lowest layer of the OSI model, the physical layer defines the electrical functionality required to send and receive bits over a given physical medium. Specifications that define the voltage levels and physical components of a network are defined at the physical layer. Protocols are not specified at the physical layer, because they are implemented as software. Examples of the standards for sending data over the physical medium are Ethernet (the most widely used standard in networking), Token Ring (IBM's proprietary networking topology), and FDDI (a standard for fiber optic networks, commonly used as a backbone).

For more information, see Chapter 4 of *CCNA Routing and Switching Exam Cram, Second Edition.*

Question 46

The correct answers are a, b, and c. IGRP is a distance vector protocol that is proprietary to Cisco and was designed to address the shortcomings of RIP. IGRP has a maximum hop count of 255 and can make more informed decisions when selecting routes. IGRP considers hop count, delay, bandwidth, and reliability when determining the best path, making it a more efficient routing protocol than RIP. Answer d is incorrect, because the maximum hop count used in IGRP is 255. Answer e is incorrect, because the maximum hop count used in RIP is 15. Answer f is incorrect because both RIP and IGRP are distance vector routing protocols. Answer g is incorrect, because RIP only uses hop count to determine the best path.

RIP is a dynamic routing protocol used to automatically update entries in a routing table. RIP is a distance vector routing protocol. RIP routers will broadcast the contents of their routing table every 30 seconds. This broadcast traffic does not make RIP very efficient for large networks. To select the best route, a RIP router selects the path with the fewest number of hops. (One hop is the distance from one router to another.) The maximum hop count used in RIP is 15. Any destination that requires more than 15 hops is considered unreachable by RIP.

For more information, see Chapter 11 of *CCNA Routing and Switching Exam Cram, Second Edition.*

Question 47

The correct answer is d. The **IP ROUTE 190.190.0.0 255.255.0.0 10.11.12.1** command will create a static route in the router's routing table that will send all packets destined for network **190.190.0.0 255.255.0.0** to router **10.11.12.1.** Answer a is incorrect, because the **IP ROUTE** command does not have a hyphen between **IP** and **ROUTE.** Answer b is incorrect, because **ROUTE** is not the command used to create a static route. Answer d is incorrect, because the **IP ROUTE** command specifies the source network first, then the destination router. A routing table can be of two basic types: static and dynamic. A static routing table is one that has its entries entered manually by an operator. On a Cisco router, a static route can be entered using the **IP ROUTE** command. The syntax for the **IP ROUTE** command is as follows:

```
IP ROUTE [network] [subnet mask] [IP address] [distance]
```

➤ *IP ROUTE*—This command signifies this will be a static entry in the routing table.

➤ *[network]*—This is the destination network ID.

➤ *[subnet mask]*—This is the subnet mask of the destination network.

➤ *IP ADDRESS*—This is the IP address of the router that will receive all packets that have the address specified in the **network** statement.

➤ *[distance]*—This is an optional parameter and is the administrative distance of this route, which tells the router the relative importance of this route. If a router has multiple entries in its routing table for the same network ID, it will use the route with the lowest administrative distance. The default administrative distance for a static route is 1. (The number 1 is the second highest priority for administrative distance; 0 is the highest and signifies a directly connected interface.)

For more information, see Chapter 11 of *CCNA Routing and Switching Exam Cram, Second Edition.*

Question 48

The correct answer is c. The correct password to enter in this scenario is "KATE", because it is the enable secret password. Answer a is incorrect, because both the enable and the enable secret passwords are used to control access to privileged mode. Answers b and c are incorrect, because the enable secret password will take precedence over the enable password if both are enabled. The enable password is used to control access to the privileged mode of the router. If an enable password is specified, it must be entered after the **ENABLE** command to successfully access privileged mode. To configure the enable password, use the **ENABLE PASSWORD** *[password]* command. The enable secret password is used to control access to privileged mode, similar to the enable password. The difference between enable secret and enable password is that the enable secret password will be encrypted for additional security.

For more information, see Chapter 8 of *CCNA Routing and Switching Exam Cram, Second Edition.*

Question 49

The correct answer is d. **CONFIGURE MEMORY (CONFIG MEM)** is used to execute the configuration stored in NVRAM and will copy the startup configuration to the running configuration. Answer a is incorrect, because the **CONFIGURE TERMINAL** command is used to enter commands into the router from the console or through Telnet. Answer b is incorrect, because the **CONFIGURE NETWORK** command is used to copy the configuration file from a TFTP server into the router's RAM. Answer c is incorrect, because the **CONFIGURE OVERWRITE-NETWORK** command is used to copy a configuration file into NVRAM from a TFTP server. The **CONFIGURE** command, executed in privileged mode, is used to enter the configuration mode of a Cisco router. The four parameters used with the **CONFIGURE** command are **TERMINAL**, **MEMORY**, **OVERWRITE-NETWORK**, and **NETWORK**. The following list outlines the functions of these parameters:

➤ *CONFIGURE TERMINAL (CONFIG T)*—This command is used to enter configuration commands into the router from the console port or through Telnet.

➤ *CONFIGURE NETWORK (CONFIG NET)*—This command is used to copy the configuration file from a TFTP server into the router's RAM.

➤ *CONFIGURE OVERWRITE-NETWORK (CONFIG O)*—This command is used to copy a configuration file into NVRAM from a TFTP server. Use of the **CONFIGURE OVERWRITE-NETWORK** will not alter the running configuration.

➤ *CONFIGURE MEMORY (CONFIG MEM)*—This command is used to execute the configuration stored in NVRAM. It will copy the startup configuration to the running configuration.

For more information, see Chapter 8 of *CCNA Routing and Switching Exam Cram, Second Edition.*

Question 50

The correct answer is c. The **SETUP** command will execute the System Configuration dialog, which can be used to modify the configuration used by the router. Answer a is incorrect, because the **SETUP** command must be run from privileged mode. In privileged mode, the router prompt is followed by a pound sign (**ROUTER#**). In user mode, the router prompt is followed by an angle bracket (**ROUTER>**). Answers b and d are incorrect, because **INSTALL** and **SYS-CONFIG** are not valid Cisco IOS commands.

For more information, see Chapter 8 of *CCNA Routing and Switching Exam Cram, Second Edition.*

Question 51

The correct answer is b. The **ROUTER IGRP 15 NETWORK 200.200.200.0** command will enable IGRP on the router and have it advertise network **200.200.200.0** in autonomous system number **15**. Answer a is incorrect, because the autonomous system number is supposed to be entered after the **ROUTER IGRP** command. Answer c is incorrect, because the **ROUTER IGRP** command does not have a hyphen between **ROUTER** and **IGRP**. Answer d is incorrect, because the **ROUTER IGRP** command does not use **AS** after **ROUTER IGRP**. IGRP is a distance vector protocol that is proprietary to Cisco and was designed to address the shortcomings of RIP. IGRP has a maximum hop count of 255 and can make more informed decisions when selecting routes. IGRP considers hop count, delay, bandwidth, and reliability when determining the best path, making it a more efficient routing protocol than RIP. To enable IGRP on a Cisco router, use the **ROUTER IGRP** command. The syntax for the **ROUTER IGRP** command is as follows:

```
ROUTER IGRP [autonomous system number] NETWORK [network ID]
```

➤ *ROUTER IGRP*—This command enables IGRP on the router.

➤ *[autonomous system number]*—This parameter identifies the group of IGRP routers that this router will communicate with. All routers must have the same autonomous system number to communicate.

➤ *NETWORK [network ID]*—This command specifies the network ID that this router will advertise.

For more information, see Chapter 11 of *CCNA Routing and Switching Exam Cram, Second Edition.*

Question 52

The correct answer is b. Privileged mode is used to modify and view the configuration of the router, as well as set IOS parameters. The Cisco IOS uses a command interpreter known as **EXEC**, which contains two primary modes: user mode and privileged mode. User mode is used to display basic router system information and to connect to remote devices. Answers a and d are incorrect, because the commands in user mode are limited in capability. When you first log into a router, you are placed in user mode, in which the router prompt is followed by an angle bracket **(ROUTER>)**. Privileged mode also contains the **CONFIGURE** command, which is used to access other configuration modes, such as global configuration mode and interface mode. In privileged mode, the router prompt is followed by a pound sign **(ROUTER#)**. To enter privileged mode, enter the **ENABLE** command from user mode.

For more information, see Chapter 8 of *CCNA Routing and Switching Exam Cram, Second Edition.*

Question 53

The correct answer is c. Because the store-and-forward method of switching is the only method that performs error checking of packets before they are sent, all of the other answers are incorrect. A switch is a data-link layer connectivity device used to segment a network. A switch resides at the data-link layer of the OSI model and can filter traffic based on the data-link address of a packet or MAC address. (The MAC address is the address burned into the network adapter card by the manufacturer.)

Cisco switches employ two basic methods of forwarding packets: cut-through and store-and-forward. Store-and-forward switches will copy an incoming packet to the local buffer and perform error checking on the packet before sending it on to its destination. If the packet contains an error, it is discarded. Cut-through switches will read the destination address of an incoming packet and immediately search its switch table for a destination port. Cut-through switches will not perform any error checking of the packet. The fact that cut-through switches do not perform error checking or copy the packet to their buffer before processing means they experience less latency (delay) than store-and-forward switches.

For more information, see Chapter 13 of *CCNA Routing and Switching Exam Cram, Second Edition.*

Question 54

The correct answers are a and d. Network layer protocols include IP (part of the TCP/IP protocol suite) and Internetwork Packet Exchange (IPX, part of the IPX/SPX protocol suite). Both IP and IPX are used to provide a logical address that is used to identify a host on the network. Answers b and c are incorrect, because TCP (part of the TCP/IP protocol suite) and Sequenced Packet Exchange (SPX, part of the IPX/SPX protocol suite) are transport layer protocols and do not provide logical network layer addresses.

For more information, see Chapter 4 of *CCNA Routing and Switching Exam Cram, Second Edition.*

Question 55

The correct answer is c. Key components of WAN communication are data terminal equipment (DTE) and data circuit-terminating equipment (DCE) devices. Answers a and b are incorrect, because a DTE is usually the router that acts as the gateway to the local area network (LAN). Answer d is incorrect, because a DCE are the devices between the LAN and the DTE that perform the data translation between the LAN and the DTE. Common DCE devices are modems (used when communicating over analog phone lines), and a channel service unit/data service unit (CSU/DSU, used when communication is over digital lines).

For more information, see Chapter 16 of *CCNA Routing and Switching Exam Cram, Second Edition.*

Question 56

The correct answer is c. The access list in the question is a standard IP access list, which can be easily identified by referring to the number after the AC-CESS-LIST command. Answer a is incorrect, because the range of a standard IPX access list is 800–899. Answer b is incorrect, because the range of an extended IP access list is 100–199. Answer d is incorrect, because the range of an extended IPX access list is 900–999. The five major types of access lists utilize the following ranges of numbers:

➤ *IP Standard Access List*—1–99

➤ *IP Extended Access List*—100–199

➤ *IPX Standard Access List*—800–899

➤ *IPX Extended Access List*—900–999

➤ *IPX SAP Access List*—1000–1099

For more information, see Chapter 13 of *CCNA Routing and Switching Exam Cram, Second Edition.*

Question 57

The correct answer is c. Answers a and d are incorrect, because this is a Class b IP address. Answer b is incorrect, because the subnet mask is 255.255.248.0. When configuring IP on a Cisco router interface, you must specify the IP address for the interface and the number of bits used in the subnet mask. When an IP address is specified for the interface, the router will identify the address by the first octet rule as either a Class A, B, or C address. (Class A=1–126, Class B=128–191, and Class C=192–223.) The default subnet mask for the address class is assumed to be present by the router. (Class A=255.0.0.0, Class B=255.255.0.0, and Class C=255.255.255.0.)

The second step to configuring IP on a router interface involves specifying the number of bits in the subnet field. If this number is 0, the default subnet mask is applied based on the address class. The number of bits specified in the subnet field will be applied by the router to the host portion of the IP address. In this question, a Class B address was entered for the interface (131.200.15.2). The router then assumed a subnet mask of 255.255.0.0, the default subnet mask for a Class B address. The number of bits in the subnet field was specified as 5, which was interpreted by the router as 5 bits in the *third* octet of the IP address and is applied as 255.255.248.0. (Five bits in binary=11111000, which is the decimal equivalent to 248.)

For more information, see Chapter 8 of *CCNA Routing and Switching Exam Cram, Second Edition.*

Question 58

The correct answer is c. Basic Rate ISDN (also known as 2B+1D) comprises two B channels that are 64Kbps each and one 16Kbps D channel. Answers a, b, and c are incorrect, because Basic Rate ISDN (also known as 2B+1D) comprises two B channels that are 64Kbps each and one 16Kbps D channel. ISDN is a communications standard that uses digital telephone lines to transmit voice, data, and video, and it is a dial-up service that is used on demand. The bandwidth provided by ISDN can be divided into two categories, PRI and BRI. Basic Rate ISDN (also known as 2B+1D) comprises two B channels that are 64Kbps each and one 16Kbps D channel. Basic Rate ISDN uses the two B channels to transmit data and uses the 16Kbps D channel for control and signaling purposes. Primary Rate ISDN (also known as 23B+1D) comprises 23 B channels that are 64Kbps each and one that is a 64Kbps D channel.

For more information, see Chapter 11 of *CCNA Routing and Switching Exam Cram, Second Edition.*

Question 59

The correct answer is d. A Cisco router's startup sequence is divided into three basic operations. The POST executes, the operating system is loaded, and, finally, the startup configuration is loaded into RAM. The following steps detail the startup sequence:

1. The bootstrap program in ROM executes the POST, which performs a basic hardware level check.

2. The Cisco IOS (operating systems software) is loaded into memory per instructions from the boot system command. The IOS can be loaded from Flash, ROM, or a network location (TFTP server).

3. The router's configuration file is loaded into memory from NVRAM. If no configuration file is found, the setup program automatically initiates the dialog necessary to create a configuration file.

Answers a and c are incorrect, because the router's operating system software must load before any configuration files are loaded. Answer b is incorrect, because the bootstrap program in ROM executes the POST in step one of the start-up process.

For more information, see Chapter 4 of *CCNA Routing and Switching Exam Cram, Second Edition.*

Question 60

The correct answers are b and d. A network address or logical address is an address that resides at the network layer of the OSI model. Answer a is incorrect, because network addresses are hierarchical in nature. Answer d is incorrect, because flat addressing makes it more difficult to logically group complex networks. A hierarchical addressing scheme uses logically structured addresses to provide a more organized environment. Examples of hierarchical network addresses are IP and IPX. An IP address comprises a network ID, which identifies the network a host belongs to and a host ID, which is unique to that host. An IPX address utilizes a similar network ID and host ID format in addressing. The hierarchical addressing of IP and IPX enable complex networks to be logically grouped and organized.

A data-link address is also known as a physical address, hardware address, or, more commonly, a MAC address. (A MAC address resides at the MAC sublayer of the data-link layer.) The MAC address is the address "burned" into every network adapter card by the manufacturer. This address is considered a "flat" address, because no logical arrangement of these addresses is on a network. A flat addressing scheme simply gives each member a unique identifier that is associated with them.

For more information, see Chapter 11 of *CCNA Routing and Switching Exam Cram, Second Edition*.

Question 61

The correct answer is d. The access list **ACCESS LIST 101 PERMIT TCP 200.200.200.0 0.0.0.255 151.120.0.0 0.0.255.255 EQ 21** will allow all hosts on network **200.200.200.0** to have FTP access to network **151.120.0.0**. Answer a is incorrect, because the port number used is 23, which specifies Telnet traffic, not FTP traffic. FTP uses port 21. Answer b is incorrect, because the **ACCESS-LIST** command does not begin with "**IP**". Answer c is incorrect, because the access list number is 99, which is used for standard access lists, not extended. Extended access lists use numbers between 100 and 199.

The syntax for creating an extended IP access list is as follows:

```
ACCESS-LIST [access-list-number] [permit|deny] [protocol] [source]
    [wildcard mask] [destination] [wildmask] [operator][port]
```

➤ *ACCESS-LIST*—This is the command to denote an access list.

➤ *[access-list-number]*—This can be a number between 100 and 199 and is used to denote an extended IP access list.

➤ *[permit|deny]*—This will allow (permit) or disallow (deny) traffic specified in this access list.

➤ *[protocol]*—This is used to specify the protocol that will be filtered in this access list. Common values are TCP, UDP, and IP. (Use of IP will denote all IP protocols.)

➤ *[source]*—This specifies the source IP addressing information.

➤ *[wildcard mask]*—This can be optionally applied to further define the source. A wildmask can be used to control access to an entire IP network ID rather than a single IP address. A wildcard mask will use the number 255 to mean "any" and the number 0 to mean "must match exactly." The terms *host* and *any* may also be used here to more quickly specify a single host or an entire network. The **HOST** keyword is the same as the wildcard mask 0.0.0.0; the **ANY** keyword is the same as the wildcard mask 255.255.255.255.

➤ *[destination]*—This specifies the destination IP addressing information.

➤ *[wildmask]*—This can be optionally applied to further define the destination IP addressing.

➤ *[operator]*—This can be optionally applied to define how to interpret the value entered in the **[port]** section of the access list. Common values are equal to the port specified **EQ, LT**, and **GT**.

➤ *[port]*—This specifies the port number this access list will act on. Common port numbers include 21 (FTP), 23 (TELNET), 25 (SMTP), 53 (DNS), 69 (TFTP), and 80 (HTTP).

For more information, see Chapter 13 of *CCNA Routing and Switching Exam Cram, Second Edition.*

Question 62

The correct answer is d. The IP address range, 129.23.98.33 to 129.23.98.62, are valid host IDs for the nodes in this subnet. All of the other answers are incorrect address ranges for this subnet. When subnetting IP addresses, host IDs are grouped together according to the incremental value of the subnet mask in use. The first address in this identifies the subnet (subnet ID) and is not used for a host ID. The last address in this range is used by the router as the broadcast address. In this question, the subnet mask used is 224, which will divide subnets in groups of 32 (129.23.98.32–129.23.98.63, 129.23.98.64–129.23.98.95, 129.23.98.96–129.23.98.127, and so forth). The broadcast address is specified as 129.23.98.63, which places this address in subnet 129.23.98.32–129.23.98.63. The first address is used as the subnet ID and the last address is used for the broadcast address, making the valid range of host IDs 129.23.98.33–129.23.98.62. The following list details the incremental values of IP subnet masks:

➤ *Subnet Mask 192*—Increments in 64

➤ *Subnet Mask 224*—Increments in 32

➤ *Subnet Mask 240*—Increments in 16

➤ *Subnet Mask 248*—Increments in 8

➤ *Subnet Mask 252*—Increments in 4

➤ *Subnet Mask 254*—Increments in 2

➤ *Subnet Mask 255*—Increments in 1

For more information, see Chapter 8 of *CCNA Routing and Switching Exam Cram, Second Edition.*

Question 63

The correct answer is a. ROM is a physical chip installed on a router's motherboard that contains the bootstrap program, the POST, and the operating system software (Cisco IOS). Answer b is incorrect, because NVRAM is used by Cisco routers to store the startup configuration Answer c is incorrect, because the Flash is an erasable, programmable memory area that contains the Cisco operating system software (IOS). Answer d is incorrect, because RAM is used as the router's main working area and contains the running configuration.

For more information, see Chapter 4 of *CCNA Routing and Switching Exam Cram, Second Edition.*

Question 64

The correct answer is a. The **LINE VTY 0 4/LOGIN/PASSWORD JETS** commands will require the password "JETS" to be entered by users connecting through Telnet. The virtual terminal password is used to control remote Telnet access to a router. Setting this password will require all users that Telnet into the router to provide this password for access. Answer b is incorrect, because the **LINE CON** command is used to set a console password, not a virtual terminal password. Answer c is incorrect, because **LOGIN JETS** is not a valid Cisco IOS command. Answer d is incorrect, because **LINE AUX** will set an auxiliary password, not a virtual terminal password.

For more information, see Chapter 8 of *CCNA Routing and Switching Exam Cram, Second Edition.*

Question 65

The correct answer is a. The **IPX OUTPUT-SAP-FILTER 1000** is the correct command to use in this question to apply the access list **1000** to the outgoing interface of the router. Answer b is incorrect, because the command must begin with **IPX**. Answer c is incorrect, because no hyphens are between **IPX**, **SAP**, and **FILTER**. Answer d is incorrect, because this command is used to apply a standard or extended IPX access list, not an **IPX SAP** access list. For an **IPX SAP** access list to filter traffic, it must be applied to an interface. This is done using the **IPX-OUTPUT/INPUT-SAP-FILTER** command. This command must be entered in interface configuration mode. To access interface configuration mode, use the **INTERFACE** *[interface-number]* command from global configuration mode. When the router is in interface configuration mode, the router prompt displays "**config-if**" in parentheses, followed by a pound sign (**(ROUTER(config-if)#**). The **IPX-OUTPUT/INPUT-SAP-FILTER** command has the following syntax:

```
IPX[output|input]-SAP-FILTER [access list number]
```

The **[output|input]** parameter specifies where the access list will be applied on the router. **OUTPUT** will cause the router to apply the access list to all outgoing packets, and **INPUT** will cause the router to apply the access list to all incoming packets.

For more information, see Chapter 13 of *CCNA Routing and Switching Exam Cram, Second Edition.*

Question 66

The correct answers are a, d, and f. IP address 222.111.101.200 is a Class C address with a network ID of 222.111.101.0 and a host ID of 200. Answer b is incorrect, because this is a Class c address. Answer c is incorrect, because the network ID is 222.111.101.0. Answer e is incorrect, because the host ID is 200. An IP address is a 32-bit network layer address that comprises a network ID and a host ID. TCP/IP addresses are divided into three main classes: Class A, Class B, and Class C. The first octet rule specifies that an IP address can be identified by class according to the number in the first octet. The address range for each address class is as follows:

➤ *Class A*—IP addresses with numbers from 1 to 126 in the first octet. (Note: 127 is not used as a network ID in IP; it is reserved for localhost loopback testing.)

➤ *Class B*—IP addresses with numbers from 128 to 191 in the first octet.

➤ *Class C*—IP addresses with numbers from 192 to 223 in the first octet.

When looking at any IP address, it is the number in the *first* octet that determines the address class of that address.

➤ The IP address 3.100.32.7 is a Class A address, because it *begins* with 3, a number between 1 and 126.

➤ The IP address 145.100.32.7 is a Class B address, because it *begins* with 145, a number between 128 and 191.

➤ The IP address 210.100.32.7 is a Class C address, because it *begins* with 210, a number between 192 and 223.

An IP address is made up of a network ID and a host ID. The network ID identifies the network a node is on, and the host ID identifies the individual node. An IP address is specified using four individual numbers separated by three periods. Each one of these "sections" is known as an octet (because it contains 8 bits). Each address class uses different combinations of octets to specify a network ID and a host ID. The configuration is as follows:

➤ *Class A*—This class uses the first octet for the network ID and the last three octets for the host ID.

➤ IP address 2.3.45.9 is a Class A address (it begins with 2, a number between 1 and 126). The network ID of this IP address is 2, and the host ID is 3.45.9, because a Class A address uses the first octet for the network ID and the last three octets for the host ID.

➤ *Class B*—This uses the first two octets for the network ID and the last two octets for the host ID.

➤ IP address 167.211.32.5 is a Class B address (it begins with 167, a number between 128 and 191). The network ID of this IP address is 167.211 and the host ID is 32.5, because a Class B address uses the first two octets for the network ID and the last two octets for the host ID.

➤ *Class C*—This uses the first three octets for the network ID and the last octet for the host ID.

➤ IP address 200.3.11.75 is a Class C address (it begins with 200, a number between 192 and 223). The network ID of this IP address is 200.3.11 and the host ID is 75, because a Class C address uses the first three octets for the network ID and the last octet for the host ID.

For more information, see Chapter 8 of *CCNA Routing and Switching Exam Cram, Second Edition.*

Question 67

The correct answer is c. The OSI model defines standards used in networking and comprises a seven-layer model. The session layer of the OSI model is primarily used to coordinate communication between nodes, as well as creating, maintaining, and ending a communication session. Since no other layer of the OSI model coordinates communication, all of the other answers are incorrect.

The following list outlines the seven layers of the OSI model and their functions:

➤ *Application*—The "window" to networking used by programs, the application layer is responsible for:

➤ Verifying that the appropriate resources are present to initiate a connection with the destination node.

➤ Verifying the identity of the destination node.

Application layer protocols include Telnet, FTP, SMTP, WWW, EDI, and WAIS.

➤ *Presentation*—Essentially a translator, the presentation layer is responsible for:

➤ Translating text and data syntax, such as EBCDIC, used in IBM systems, to ASCII, used in PCs and most computer systems.

➤ Using ASN 1 to perform data translation.

Presentation layer protocols include PICT, TIFF, JPEG, MIDI, MPEG, and Quick Time.

➤ *Session*—Used to coordinate communication between nodes, the session layer is responsible for:

➤ Creating, maintaining, and ending a communication session.

➤ Coordinating service requests and responses that occur between nodes across the network.

Session layer protocols include NFS, used by SUN Microsystems and Unix with TCP/IP; SQL, used to define database information requests; RPCs, used in Microsoft network communication; X Window, used by Unix terminals; ASP, used by Apple computers; and DNA SCP, used by IBM.

➤ *Transport*—Reliable end-to-end communication is the primary function of the transport layer. Its many responsibilities include:

➤ Ensuring flow control so the amount of data being sent from one node will not overwhelm the destination node.

➤ Ensuring the ability of multiple applications to utilize a single transport (multiplexing).

➤ Ensuring the reliable transfer of data. The transport layer implements connection-oriented services between nodes, which utilize a three-way handshake (synchronization, acknowledgment, and data-transfer) to efficiently transfer data.

➤ Ensuring positive acknowledgment, or the process of a node waiting for an acknowledgment from the destination node prior to sending data.

➤ Windowing, a form of flow control that specifies how much data will be transferred between acknowledgments.

Transport layer protocols include TCP, part of the TCP/IP protocol suite, and SPX, part of the IPX/SPX protocol suite.

➤ *Network*—Selecting the appropriate path a packet should take to get to its intended destination is the function of the network layer. It is responsible for routing, which is the process of using a network layer address to determine the best path a packet will travel to its destination. Network layer protocols include IP and IPX, part of the IPX/SPX protocol suite.

➤ *Data Link*—Divided into two sublayers, MAC and LLC, the data-link layer is responsible for:

➤ Preparing data from upper layers to be transmitted over the physical medium by encapsulating upper layer data into frames. This frame includes the source and destination of MAC addresses in the frame header.

➤ Converting data into bits, so it can be transmitted by the physical layer.

➤ Adding a CRC to the end of a frame, which is used for error checking at the data-link layer.

Data-link protocols include HDLC, the Cisco default encapsulation for serial connections; SDLC, used in IBM networks; LAPB, used with X.25; X.25, a packet switching network; SLIP, an older TCP/IP dial-up protocol; PPP, a newer dial-up protocol; ISDN, a dial-up digital service; and frame relay, a packet switching network.

➤ *Physical*—The lowest layer of the OSI model, the physical layer defines the electrical functionality required to send and receive bits over a given physical medium. Specifications that define the voltage levels and physical components of a network are defined at the physical layer. Protocols are not specified at the physical layer, because they are implemented as software. Examples of the standards for sending data over the physical medium are Ethernet (the most widely used standard in networking), Token Ring (IBM's proprietary networking topology), and FDDI (a standard for fiber optic networks, commonly used as a backbone).

For more information, see Chapter 4 of *CCNA Routing and Switching Exam Cram, Second Edition*.

Question 68

The correct answers are a and b. The **SHOW STARTUP-CONFIG** command can be used in privileged mode to display the startup configuration file stored in NVRAM and can be abbreviated as **SH STAR**. Answers c, e, and f are incorrect, because the command is entered in user mode. In this mode, the router prompt is followed by an angle bracket **(ROUTER>)**. The **SHOW STARTUP-CONFIG** command cannot be used in either user or global configuration modes. In privileged mode, the router prompt is followed by a pound sign **(ROUTER#)**. Answer g is incorrect, because **DISPLAY STARTUP** is not a valid Cisco IOS command.

For more information, see Chapter 8 of *CCNA Routing and Switching Exam Cram, Second Edition*.

Question 69

The correct answer is d. **CONFIGURE MEMORY (CONFIG MEM)** is used to execute the configuration stored in NVRAM and will copy the startup configuration to the running configuration. Answer a is incorrect, because the **CONFIGURE TERMINAL** command is used to enter configuration commands into the router from the console port or through Telnet. Answer b is incorrect, because the **CONFIGURE OVERWRITE-NETWORK** command is used to copy a configuration file into NVRAM from a TFTP server. Answer c is incorrect, because the **CONFIGURE NETWORK** command is used to copy the configuration file from a TFTP server into the router's RAM. The **CONFIGURE** command, executed in privileged mode, is used to enter the configuration mode of a Cisco router. The four parameters used with the

CONFIGURE command are **TERMINAL, MEMORY, OVERWRITE-NETWORK,** and **NETWORK.** The following list outlines the functions of these parameters:

➤ *CONFIGURE TERMINAL (CONFIG T)*—This is used to enter configuration commands into the router from the console port or through Telnet.

➤ *CONFIGURE NETWORK (CONFIG NET)*—This is used to copy the configuration file from a TFTP server into the router's RAM.

➤ *CONFIGURE OVERWRITE-NETWORK (CONFIG O)*—This is used to copy a configuration file into NVRAM from a TFTP server. Use of the **CONFIGURE OVERWRITE-NETWORK** will not alter the running configuration.

➤ *CONFIGURE MEMORY (CONFIG MEM)*—This is used to execute the configuration stored in NVRAM. It will copy the startup configuration to the running configuration.

For more information, see Chapter 8 of *CCNA Routing and Switching Exam Cram, Second Edition.*

Question 70

The correct answer is d. Pressing the Control key along with the letter *N* will return to more recent commands in the command history, after recalling previous commands. Configuring a Cisco router can sometimes involve long, detailed command lines that can be slow to navigate. For this reason, the Cisco IOS includes a number of shortcuts for navigating them. The following list details some of the more common methods:

➤ *Ctrl+A*—Pressing the Control key along with the letter *A* will move the cursor to the beginning of the command line.

➤ *Ctrl+E*—Pressing the Control key along with the letter *E* will move the cursor to the end of the command line.

➤ *Ctrl+P*—Pressing the Control key along with the letter *P* will recall previous commands in the command history. (The up arrow key will do this as well.)

➤ *Ctrl+N*—Pressing the Control key along with the letter *N* will return to more recent commands in the command history after recalling previous commands. (The down arrow key will do this as well.)

➤ *Esc+B*—Pressing the Escape key along with the letter *B* will move the cursor back one word.

➤ *Esc+F*—Pressing the Escape key along with the letter *F* will move the cursor forward one word.

➤ *Left and Right Arrow Keys*—These arrow keys will move the cursor one character left and right, respectively.

➤ *Tab*—Pressing the Tab key will complete an entry typed at the router prompt.

For more information, see Chapter 8 of *CCNA Routing and Switching Exam Cram, Second Edition*.

The Coriolis Exam Cram Personal Trainer

An exciting new category in certification training products

The Exam Cram Personal trainer is the first certification-specific testing product that completely links learning with testing to:

- **Increase your comprehension**
- **Decrease the time it takes you to learn**

No system blends learning content with test questions as effectively as the Exam Cram Personal Trainer.

Only the Exam Cram Personal Trainer offers this much power at this price.

Its unique Personalized Test Engine provides a real-time test environment and an authentic representation of what you will encounter during your actual certification exams.

Much More Than Just Another CBT!

Most current CBT learning systems offer simple review questions at the end of a chapter with an overall test at the end of the course, with no links back to the lessons. But Exam Cram Personal Trainer takes learning to a higher level.

Its four main components are:

- The complete text of an Exam Cram study guide in an HTML format,
- A Personalized Practice Test Engine with multiple test methods

Adaptive:	25-35 questions
Fixed-length:	Four unique exams on critical areas
Random:	Each randomly generated test is unique
Test All:	Drawn from the complete database of questions
Topic:	Organized by Exam Cram chapters
Review:	Questions with answers are presented

Scenario-based questions: Just like the real thing

- A database of nearly 300 questions linked directly to an Exam Cram chapter
- Over two hours of Exam Cram Audio Review

Plus, additional features include:

- **Hint:** Not sure of your answer? Click Hint and the software goes to the text that covers that topic.
- **Lesson:** Still not enough detail? Click Lesson and the software goes to the beginning of the chapter.
- **Update feature:** Need even more questions? Click Update to download more questions from the Coriolis Web site.
- **Notes:** Create your own memory joggers.
- **Graphic analysis:** How did you do? View your score, the required score to pass, and other information.
- **Personalized Cram Sheet:** Print unique study information just for you.

MCSE Networking Essentials Exam Cram Personal Trainer
ISBN:1-57610-644-6

MCSE NT Server 4 Exam Cram Personal Trainer
ISBN: 1-57610-645-4

MCSE NT Server 4 in the Enterprise Exam Cram Personal Trainer
ISBN: 1-57610-646-2

MCSE NT Workstation 4 Exam Cram Personal Trainer
ISBN:1-57610-647-0

A+ Exam Cram Personal Trainer
ISBN: 1-57610-658-6

$69.99 U.S. • $104.99 Canada

Available: March 2000

The <u>Smartest</u> Way to Get Certified Just Got Smarter™

Look for All of the Exam Cram Brand Certification Study Systems

ALL NEW! Exam Cram Personal Trainer Systems

The Exam Cram Personal Trainer systems are an exciting new category in certification training products. These CD-ROM based systems offer extensive capabilities at a moderate price and are the first certification-specific testing product to completely link learning with testing.

This Exam Cram Study Guide turned interactive course lets you customize the way you learn.

Each system includes:

- A Personalized Practice Test engine with multiple test methods,
- A database of nearly 300 questions linked directly to the subject matter within the Exam Cram on which that question is based.

Exam Cram Audio Review Systems

Written and read by certification instructors, each set contains four cassettes jam-packed with the certification exam information you must have. Designed to be used on their own or as a complement to our Exam Cram Study Guides, Flash Cards, and Practice Tests.

Each system includes:

- Study preparation tips with an essential last-minute review for the exam
- Hours of lessons highlighting key terms and techniques
- A comprehensive overview of all exam objectives
- 45 minutes of review questions complete with answers and explanations

Exam Cram Flash Cards

These pocket-sized study tools are 100% focused on exams. Key questions appear on side one of each card and in-depth answers on side two. Each card features either a cross-reference to the appropriate Exam Cram Study Guide chapter or to another valuable resource. Comes with a CD-ROM featuring electronic versions of the flash cards and a complete practice exam.

Exam Cram Practice Tests

Our readers told us that extra practice exams were vital to certification success, so we created the perfect companion book for certification study material.

Each book contains:

- Several practice exams
- Electronic versions of practice exams on the accompanying CD-ROM presented in an interactive format enabling practice in an environment similar to that of the actual exam
- Each practice question is followed by the corresponding answer (why the right answers are right and the wrong answers are wrong)
- References to the Exam Cram Study Guide chapter or other resource for that topic

What's On The CD-ROM

The *CCNA Routing and Switching Practice Tests Exam Cram's* companion CD-ROM contains elements specifically selected to enhance the usefulness of this book, including:

➤ The testing system for CCNA Routing and Switching Exam, which includes 70 questions. You can choose from numerous testing formats, including Fixed-Length, Random, Test All, and Review. You also have the option to save a test and go back to it later so you can take the same test over or so you can review the questions you got wrong.

System Requirements

Software:

➤ Your operating system must be Windows 95, 98, NT4, or higher.

➤ A Cisco Internetworking device running Cisco IOS software version 11.1 or later will allow you to complete the exercises and labs contained in the book but is not required to take the exams contained on the CD-ROM.

Hardware

➤ An Intel Pentium, AMD, or comparable 100MHz processor or higher is recommended for best results.

➤ 32MB of RAM is the minimum memory requirement.

➤ Available disk storage space of at least 10MB is recommended.